Liquid Marbles

Liquid Marbles

Formation, Characterization, and Applications

Andrew Terhemen Tyowua

CRC Press
Taylor & Francis Group
Boca Raton London New York

CRC Press is an imprint of the
Taylor & Francis Group, an **informa** business

CRC Press
Taylor & Francis Group
6000 Broken Sound Parkway NW, Suite 300
Boca Raton, FL 33487-2742

First issued in paperback 2020

ISBN 13: 978-0-367-57087-3 (pbk)
ISBN 13: 978-1-138-19730-5 (hbk)

Library of Congress Cataloging-in-Publication Data

Names: Tyowua, Andrew Terhemen, author.
Title: Liquid marbles : formation, characterization, and applications / Andrew Terhemen Tyowua.
Description: Boca Raton : CRC Press, Taylor & Francis Group, 2019. | Includes bibliographical references.
Identifiers: LCCN 2018027412 | ISBN 9781138197305 (hardback : alk. paper)
Subjects: LCSH: Non-Newtonian fluids--Textbooks. | Fluid dynamics--Textbooks. | Matter--Properties--Textbooks.
Classification: LCC QA929.5 .T946 2019 | DDC 532/.05--dc23
LC record available at https://lccn.loc.gov/2018027412

Visit the Taylor & Francis Web site at
http://www.taylorandfrancis.com

and the CRC Press Web site at
http://www.crcpress.com

Dedication

For my daughter, Kumawuese Michelle, who arrived during the preparation of the book.

Contents

Foreword

Colloid and Surface Chemistry is very important as it forms the basis of many pharmaceutical, agricultural, industrial, and cosmetic formulations. Therefore, it is important to teach it to students across diverse disciplines, such as chemistry, industrial chemistry, cosmetic chemistry, agricultural chemistry, chemical engineering, soft matter physics, and pharmacy. "Liquid marbles", relatively new colloids, which are macroscopic liquid drops enwrapped in hydrophobic powdered particles, have enormous applications in these areas whose understanding provides students with modern concepts for the future. This requires reference textbooks on the subject to bring together knowledge in a coherent manner and complement original journal articles. It is for this reason that *Liquid Marbles: Formation, Characterization, and Applications* was written. Author Andrew Terhemen Tyowua has a significant knowledge of the subject from completion of his doctoral studies with a world-leading authority on foams, emulsions, and colloidal particles at interfaces and from his own original research and publications on liquid marbles.

The book is organized into five chapters. Chapter 1 talks about the concepts of interfaces and surface tension, consequences of surface tension, and measurement of surface tension. Chapter 2 is on the nature of solid surfaces and contains sections on the three-phase contact angle, contact angle and surface free energy measurement, contact angle hysteresis, the Young, Wenzel, and Cassie-Baxter wetting states, wetting and non-wetting surfaces, and powdered particles. Chapter 3 talks about sticking and non-sticking drops with Leidenfrost drops and liquid marbles as modeled examples. Chapter 4 is on the preparation, properties, and characterization of liquid marbles, while Chapter 5 is on the various applications of liquid marbles. Although not all the applications of liquid marbles are discussed, a number, sufficient for teaching a thorough introduction, is provided.

I am particularly happy with the organization of the book, which provides a logical order to read or for the subject to be taught. The worked examples, exercises, and a list of other relevant materials for further reading and a full list of references in each chapter will be very helpful to students and tutors. Finally, I congratulate the author on his new book.

Glen McHale
Northumbria University

Preface

Liquid Marbles: Formation, Characterization, and Applications has come from the teaching of Colloid and Surface Chemistry at the Department of Chemistry, Benue State University, Makurdi, Nigeria. Given that "liquid marbles" are relatively new colloids, with numerous applications, it was added to the content of our Colloid and Surface Chemistry course. Unfortunately, there was no reference textbook students can be referred to for further reading, and this book is meant to solve the problem.

The book is divided into five chapters. The concepts of "interface" and surface tension, very important in Colloid and Surface Chemistry, are given in Chapter 1. The origin of interfaces and surface tension is discussed in detail. The consequences of surface tension are stated. The chapter is closed with a detailed discussion of surface tension measurement. Chapter 2 is on the nature of solid surfaces. The chapter contains discussions on the three-phase contact angle, contact angle and surface free energy measurement, contact angle hysteresis, the Young, Wenzel and Cassie-Baxter wetting states, wetting and non-wetting surfaces, and powdered particles. The chapter ends with non-wetting powdered particles, which are very important in liquid marble formation. Chapter 3 is on sticking and non-sticking drops, typically, Leidenfrost drops and liquid marbles. Chapter 4 deals with the preparation, properties, and characterization of liquid marbles. The book ends with Chapter 5, which contains the various applications of liquid marbles. In addition, each chapter contains solved examples, exercises, a list of materials for further reading, and a full list of references. It is recommended that the book be read sequentially, beginning with Chapters 1 through 5, because the chapters are related in the order presented.

Andrew Terhemen Tyowua
Benue State University

Acknowledgments

I acknowledge with thanks

- my wife, pharmacist Maryam Terhemen, for her encouragement and support.
- my son, Aondover Terhemen, for always being by my side and diverting my attention from typesetting when necessary.
- the Vice-Chancellor, Benue State University, Makurdi, Nigeria, Professor Msugh M Kembe, for his encouragement and support.
- the Dean of Science, Benue State University, Makurdi, Nigeria, Professor Stephen G Yiase, for his encouragement and support.
- my head of department, Professor Ogbene G Igbum, for her encouragement and support.
- my doctoral supervisor, Professor Bernard P Binks, the University of Hull, UK, for his continued mentorship.
- my friend, Professor Felix E Okieimen, the University of Benin, Benin-City, Nigeria, who always inspires me.
- the anonymous reviewers whose comments have added value and quality to the book.
- my editor, Dr Barbara Knott, and her assistant, Danielle Zarfati, both of the CRC Press, for coordinating all the activities related to the publication of this book.
- all my students who have inspired me and whom I enjoy teaching.

About the Author

Andrew Terhemen Tyowua obtained BSc (First Class) in Chemistry from the Benue State University, Makurdi, Nigeria in 2009. In 2015 he was awarded a PhD in chemistry by the University of Hull, United Kingdom, for his work on "Solid Particles at Fluid Interfaces" with Professor Bernard Paul Binks. His thesis was "Solid Particles at Fluid Interfaces: Emulsions, Liquid Marbles, Dry Oil Powders, and Oil Foams". Andrew spent an additional year with Professor Binks as a postdoctoral researcher before returning to Nigeria. Currently, Andrew teaches physical chemistry at the Benue State University, Makurdi, Nigeria. His research interests are formulation science and surface engineering, and he is the founder of the Applied Colloid Science and Cosmeceutical Group of the University. Andrew has authored two textbooks, namely *Fundamentals of Practical Chemistry* and *Modern Principles of Colloid and Surface Chemistry* and more than thirty articles in reputable journals such as *Langmuir* and *Soft Matter*. Andrew has received numerous academic awards. Outstanding Performance Award (value $700) was received in 2016 from the Benue State University, Makurdi, Nigeria.

Andrew and his wife, Maryam, and they are blessed with two children, Aondover Anderson and Kumawuese Michelle. Andrew and his family are currently based in Makurdi, Nigeria.

Base Units, Derived Units, Prefixes, and Conversions

SI Base Units

	Name	Symbol
length	metre	m
mass	kilogram	kg
time	second	s
thermodynamic temperature	kelvin	K
electric current	ampere	A
amount of substance	mole	mol
luminous intensity	candela	cd

Common SI Derived Units

Derived Quantity	Name	Symbol	Symbol in Base Units
frequency	hertz	Hz	s^{-1}
area		m^2	m^2
volume		m^3	m^3
density		$kg\ m^{-3}$	$kg\ m^{-3}$
velocity		$m\ s^{-1}$	$m\ s^{-1}$
acceleration		$m\ s^{-2}$	$m\ s^{-2}$
molar volume		$m^3\ mol^{-1}$	$m^3\ mol^{-1}$
molar mass		$kg\ mol^{-1}$	$kg\ mol^{-1}$
force	newton	N	$kg\ m\ s^{-2}$
energy	joule	J	$kg\ m^2\ s^{-2}$
pressure	pascal	$Pa = N\ m^{-2}$	$kg\ m^{-1}\ s^{-2}$
dynamic and shear viscosity		$Pa\ s = N\ s\ m^{-2}$	$kg\ m^{-1}\ s^{-1}$
surface tension		$N\ m^{-1}$	$kg\ s^{-2}$
heat capacity		$J\ K^{-1}$	$kg\ m^2\ s^{-2}\ K^{-1}$
molar heat capacity		$J\ K^{-1}\ mol^{-1}$	$kg\ m^2\ s^{-2}\ K^{-1}\ mol^{-1}$
latent heat of evaporation		$J\ kg^{-1}$	$m^2\ s^{-2}$
electric charge	coulomb	C	$A\ s$
electric potential	volt	$V = J\ C^{-1}$	$kg\ m^2\ s^{-3}\ A^{-1}$
power, radiant flux	watt	$W = J\ s^{-1}$	$kg\ m^2\ s^{-3}$
thermal conductivity		$W\ m^{-1}\ K^{-1}$	$kg\ m\ s^{-3}\ K^{-1}$
Celsius temperature	degree Celsius	$°C$	

SI Prefixes

Submultiple	Name	Symbol
10^{-1}	deci	d
10^{-2}	centi	c
10^{-3}	milli	m
10^{-6}	micro	μ
10^{-9}	nano	n
10^{-12}	pico	p
10^{-15}	femto	f
10^{-18}	atto	a
10^{-21}	zepto	z
10^{-24}	yocto	y

Multiple	Name	Symbol
10^{1}	deca	da
10^{2}	hecto	h
10^{3}	kilo	k
10^{6}	mega	M
10^{9}	giga	G
10^{12}	tera	T
10^{15}	peta	P
10^{18}	exa	E
10^{21}	zetta	Z
10^{24}	yotta	Y

Conversion to SI Units

Examples 1

$2\,dm = 2 \times 10^{-1}\,m$

$2\,dm^2 = 2 \times (10^{-1}\,m)^2 = 2 \times 10^{-2}\,m^2$

$2\,dm^3 = 2 \times (10^{-1}\,m)^3 = 2 \times 10^{-3}\,m^3$

$2\,hm = 2 \times 10^{2}\,m$

$2\,cm^2 = 2 \times (10^{-2}\,m)^2 = 2 \times 10^{-4}\,m^2$

$2\,cm^3 = 2 \times (10^{-2}\,m)^3 = 2 \times 10^{-6}\,m^3$

$2\,km = 2 \times 10^{3}\,m$

$2\,mm^2 = 2 \times (10^{-3}\,m)^2 = 2 \times 10^{-6}\,m^2$

$2\,mm^3 = 2 \times (10^{-3}\,m)^3 = 2 \times 10^{-9}\,m^3$

Examples 2

$2\,dam = 2 \times 10^{1}\,m$

$2\,dam^2 = 2 \times (10^{1}\,m)^2 = 2 \times 10^{2}\,m^2$

$2\,dam^3 = 2 \times (10^{1}\,m)^3 = 2 \times 10^{3}\,m^3$

$2\,cm = 2 \times 10^{-2}\,m$

$2\,hm^2 = 2 \times (10^{2}\,m)^2 = 2 \times 10^{4}\,m^2$

$2\,hm^3 = 2 \times (10^{2}\,m)^3 = 2 \times 10^{6}\,m^3$

$2\,mm = 2 \times 10^{-3}\,m$

$2\,km^2 = 2 \times (10^{3}\,m)^2 = 2 \times 10^{6}\,m^2$

$2\,km^3 = 2 \times (10^{3}\,m)^3 = 2 \times 10^{9}\,m^3$

Examples 3

$$25 \text{ mN m}^{-1} = 25\left(\frac{\text{mN}}{\text{m}}\right) = 25\left(\frac{10^{-3}\text{ N}}{\text{m}}\right) = 25 \times 10^{-3}\left(\frac{\text{N}}{\text{m}}\right) = 25 \times 10^{-3} \text{ N m}^{-1}$$

$$32 \text{ g mol}^{-1} = 32\left(\frac{\text{g}}{\text{mol}}\right) = 32\left(\frac{10^{-3}\text{ kg}}{\text{mol}}\right) = 32 \times 10^{-3}\left(\frac{\text{kg}}{\text{mol}}\right) = 32 \times 10^{-3} \text{ kg mol}^{-1}$$

$$0.997 \text{ g cm}^{-3} = 0.997\left(\frac{\text{g}}{\text{cm}^3}\right) = 0.997\left[\frac{10^{-3}\text{ kg}}{(10^{-2}\text{ m})^3}\right] = 0.997\left(\frac{10^{-3}\text{ kg}}{10^{-6}\text{ m}^3}\right)$$

$$= 0.997\left(\frac{\text{kg}}{10^{-3}\text{ m}^3}\right) = 997 \text{ kg m}^{-3}$$

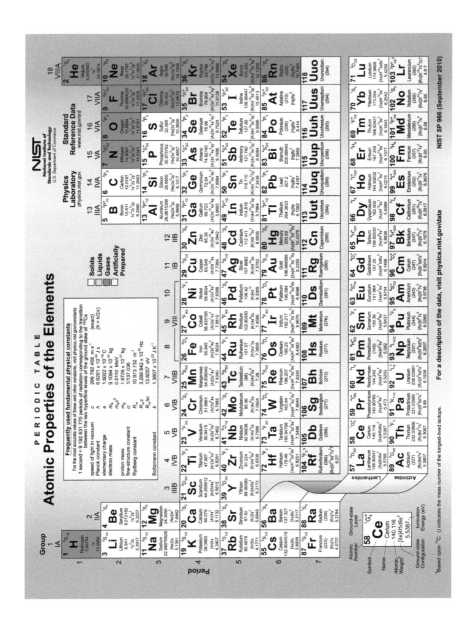

1 Interfaces and the Concept of Surface Tension

1.1 DEFINITION AND TYPES OF INTERFACES

There exists a two-dimensional plane, frontier or boundary, with no thickness, when two surfaces meet known as an "interface". A "surface", an "interphase", or "interfacial region" is a three-dimensional region, of finite thickness (few Å), where two homogeneous bulk phases meet. The properties of this region are entirely different from those of the two bulk phases. Interfaces are very common and are everywhere, *e.g.* they occur in the food we eat, bodies of living organisms, and natural and chemical environments. There are several types of interfaces, and they have been classified based on the nature of the surfaces involved in their formation. This classification yields five types of interfaces under two broad themes, namely

- Fluid interfaces: (1) liquid-air or gas and (2) liquid-liquid interfaces,
- Non-fluid or solid interfaces: (1) solid-air or gas, (2) solid-liquid and (3) solid-solid interfaces.

Because gases are miscible with one another, there are no gas-gas interfaces. Sometimes three surfaces meet in a line to form an interface. This line is called a *triple interface*.

1.2 SURFACE TENSION AND CURVED INTERFACES

1.2.1 SURFACE TENSION

Surface (or interfacial) tension is an important concept when studying fluid interfaces. Surface tension is *force per unit length* acting, perpendicularly, on an imaginary line drawn in the interface. Its SI units are Newton per metre ($N\ m^{-1}$), but it is sometimes reported in Joule per metre square ($J\ m^{-2}$) on the basis that 1 J is equivalent to 1 N m. Other units, *e.g.* dyne per centimetre, are also used. Note that 1 dyne cm^{-1} is equivalent to 1 mN m^{-1}. The surface tension (25°C) of some liquids is given in Table 1.1. A liquid surface is basically a liquid-air interface. The cohesive forces between the molecules of the liquids are greater than the adhesive forces between the air and the liquid molecules. The result is a force imbalance with a net inward force in the liquid bulk phase. This causes the liquid surface to stretch as though it were covered with an elastic membrane. This is the origin of surface tension in liquids. Surface

TABLE 1.1
Surface Tension of Some Liquids at 25°C

Liquid	γ_{la}/mN m^{-1} at 25°C
Hexane	17.89
Ethanol	21.97
Methanol	22.07
Cyclohexane	24.65
Acetone	24.02
Chloroform	26.67
Acetic acid	27.10
Toluene	27.93
Benzene	28.22
Hexadecane	27.05
Formamide	57.02
Water	71.99

Source: Jasper, J.J., *J. Phys. Chem. Ref. Data* 1, 841–1009, 1972.

tension is responsible for many observed physical phenomena. For example, due to surface tension, liquid drops minimize their surface area by taking a spherical shape (geometry of least surface area). The rising of liquids in thin capillaries once submerged in them is also a consequence of surface tension. It is also because of surface tension that insects are able to walk on the surface of water.

1.2.2 CAPILLARITY AND WICKING

A liquid will either rise against gravity or fall in a thin capillary once submerged (vertically) in it, and the phenomenon is known as *capillarity*. The rise or fall is, however, dependent on whether the cohesive forces between the liquid molecules are higher than the adhesive forces between the liquid molecules and those of the capillary. For liquids that rise in thin capillaries, the adhesive forces are higher than the cohesive forces. The converse is true for liquids that fall in capillaries. This is the reason water rises in glass capillaries while mercury falls in them. When "rising" or "falling" stops in capillaries, the liquid-air interface in them curves inwardly (concave, *e.g.* water in glass capillaries) or outwardly (convex, *e.g.* mercury in glass capillaries).

Rather than spread on porous materials, liquids invade them. This phenomenon is known as *wicking*. It is different from absorption, where the molecules of the material take in the liquid molecules. However, it is very similar to capillarity except that several pores (capillaries) are involved. Wicking is, actually, a consequence of capillary action. For example, the small pores of sponge behave like small capillaries and cause them to absorb a large amount of liquid. Textile fabrics, like the handkerchief, use capillary action to "wick" sweat away from the skin. In thin layer chromatography, solvent molecules move vertically against gravity through the pores (the gaps between the small particles) of the plate due to capillary action.

1.2.3 BUBBLES AND DROPS

To reiterate, surface tension causes fluids to minimize their surface area by taking a spherical geometry when in minute amount. This often leads to the formation of drops (or droplets), bubbles, or cavities. Drops are spheres of a liquid in equilibrium with their vapor. A bubble is either a region in which air and vapor are trapped by a thin film or a cavity full of vapor in a liquid. Ordinary bubbles have two surfaces (*i.e.* inner and outer), while cavities have only one surface. Drops and bubbles are stable (*i.e.* at equilibrium) because the tendency to decrease their surface area is balanced by the rise in internal pressure. The internal pressure, P_i, which tends to expand them, is usually greater than the outer (or external atmospheric) pressure, P_o, which tends to contract them. Consider a relatively stable one-surface air bubble, Figure 1.1a, of radius r in a liquid of surface tension γ_{la}. A small change (increment) in the radius, from r to $r + dr$, will be accompanied by a change in the surface area $d\alpha$.

$$d\alpha = 4\pi(r + dr)^2 - 4\pi r^2 = 4\pi[r^2 + 2rdr + (dr)^2] - 4\pi r^2 \tag{1.1}$$

$$= 4\pi r^2 + 8\pi rdr + 4\pi(dr)^2 - 4\pi r^2 \approx 8\pi rdr \tag{1.2}$$

In Equation (1.2), the $(dr)^2$ term is neglected since it is negligibly small. The work done dW in stretching the surface by this amount is

$$dW = \gamma_{la}d\alpha = 8\pi\gamma_{la}rdr \tag{1.3}$$

Because force × distance is work done, the force F opposing the stretching through the distance dr is $8\pi\gamma_{la}r$. The outward force is internal pressure × area, which is $4\pi r^2 P_i$, while the inward force due to the outer pressure is outer pressure × area, which is $4\pi r^2 P_o$. The total inward force is the sum of that due to the outer pressure and that due to surface tension. The bubble is at equilibrium when the outward and inward forces balance out.

$$4\pi r^2 P_i = 4\pi r^2 P_o + 8\pi\gamma_{la}r \tag{1.4}$$

Dividing Equation (1.4) throughout by $4\pi r^2$ gives the Laplace (or Young-Laplace) Equation (1.5).

$$P_i = P_o + \frac{2\gamma_{la}}{r} \text{ or } \Delta P = \frac{2\gamma_{la}}{r}; \ \Delta P = \text{ Laplace pressure} \tag{1.5}$$

In the case of a soap bubble characterized by two surfaces, Figure 1.1b, $\Delta P = 4\gamma_{la}/r$. Because $\Delta P \propto 1/r$, smaller bubbles have higher Laplace pressure than bigger ones. This is why champagne (bubbles $r < 0.1$ mm) is louder than beer (bubbles $r > 0.1$ mm) when opened. The $\Delta P \to 0$ as $r \to \infty$ (*i.e.* when the surface is flat), meaning that a pressure differential exists only across a curved interface.

Unlike in the case of a spherical interface where only one radius of curvature is required, two radii of curvature are required for a non-spherical interface, as

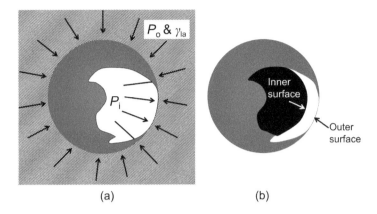

(a)   (b)

FIGURE 1.1 (a) Forces acting on a one-surface air bubble in water. The inner pressure P_i acts outwardly while the outer pressure P_o and the surface tension forces γ_{la} act inwardly. (b) Schematic representation of a soap bubble showing its two (inner and outer) surfaces.

illustrated in Figure 1.2. The two radii of curvature at a chosen point (r_A and r_B) are obtained by first taking the normal to the surface at that point followed by drawing a plane containing the normal, in an arbitrary orientation, and drawing a second plane containing the normal at right angles to the first plane. The curves formed by the intersection of the two planes with the surface are noted. Finally, the radii of the two circles drawn in the planes that have the same curvatures as these lines at the chosen point are obtained. (It is important to state that positive values are assigned to the radii of curvature, r, r_A or r_B, provided they lie in the concave side.) Therefore, the Laplace equation for the system is

$$\Delta P = \gamma_{la}\left(\frac{1}{r_A} + \frac{1}{r_B}\right) = \frac{2\gamma_{la}}{r_m}; \text{ where } \frac{1}{r_m} = \frac{1}{2}\left(\frac{1}{r_A} + \frac{1}{r_B}\right) \tag{1.6}$$

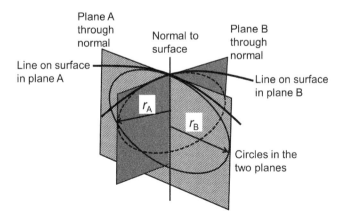

FIGURE 1.2 Schematic description of the procedure for obtaining the radii of curvature at a given point on a curved (non-spherical) interface.

with $1/r_m$ being the mean curvature (inverse of the radius). Note that Equation (1.6) is equivalent to (1.5) when $r_A = r_B$.

Example 1.1: Calculating the Laplace Pressure (or Pressure Difference) across a Bubble Surface

Calculate the pressure difference across a champagne bubble of radius 1×10^{-4} m if the surface tension of the champagne is 50 dynes cm^{-1} at 25°C.

Method

The surface tension would be converted to the N m^{-1} and because the bubble has one surface, Equation (1.5) would be used to calculate the pressure difference ΔP. Therefore, $\gamma_{la} = 5 \times 10^{-2}$ N m^{-1}. Also, note that 1 N m^{-2} = 1 Pa.

Answer

$$\Delta P = \frac{2 \times 5 \times 10^{-2} \, \text{N m}^{-1}}{1 \times 10^{-4} \, \text{m}} = 1000 \, \text{N m}^{-2} = 1 \, \text{kPa}$$

The pressure is enough to support a 10 cm column of water if its density is 1000 kg m^{-3}.

Based on Equation (1.5), bubbles and drops with different curvatures ($1/r$) will have different stabilities. Suppose that two liquid drops (of the same surface tension) of radii r_1 and r_2, where $r_1 > r_2$, are subjected to the same external pressure. The difference between the pressures P_1 and P_2 within the drops would be:

$$P_1 - P_2 = \Delta P = \frac{2\gamma_{la}}{r_1} - \frac{2\gamma_{la}}{r_2} = 2\gamma_{la}\left(\frac{1}{r_1} - \frac{1}{r_2}\right), \text{ where } P_2 > P_1 \qquad (1.7)$$

If the two drops are connected by a thin hollow tube (as shown in Figure 1.3), liquid molecules will flow *via* the tube from drop 2 to drop 1 until the small drop (2) vanishes or the pressure within both drops balances out. The same thing will be observed if two

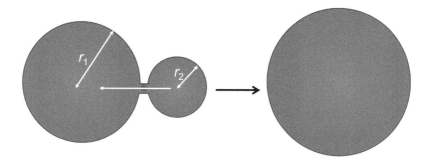

FIGURE 1.3 Two liquid drops of different sizes ($r_1 > r_2$) connected by a thin hollow tube. Because the internal pressure in the smaller is higher than that in the bigger, liquid molecules flow from the smaller drop to the bigger one until the smaller drop disappears.

air bubbles of different sizes are considered. There is no pressure gradient for the case where $r_1 = r_2$ and the phenomenon does not occur.

1.2.4 THE KELVIN EQUATION

The vapor pressure above a flat liquid surface is different from that above a curved surface. The Kelvin equation shows how the vapor pressure of a drop depends on its radius r. The equilibrium condition between the vapor phase on the outer side of the surface and the liquid phase on the inner side of the surface is that their chemical potentials μ must be equal.

$$\mu_{\text{inner}} = \mu_{\text{outer}} \Rightarrow \mu(l) = \mu(g) \tag{1.8}$$

Taking the differential of both sides of Equation (1.8) at constant temperature gives Equation (1.9), where $V_m(l)$ and $V_m(g)$ are the molar volumes of the liquid and the vapor at constant temperature, respectively.

$$d\mu(l) = V_m(l) \times dP(l); d\mu(g) = V_m(g) \times dP(g)\left(\text{since}\left(\frac{\partial \mu}{\partial P}\right)_T = V_m\right) \tag{1.9}$$

Therefore, based on Equation (1.8),

$$V_m(l)dP(l) = V_m(g)dP(g) \tag{1.10}$$

If the vapor is considered perfect, $V_m(g) = RT/P(g)$ and Equation (1.10) becomes (1.11).

$$V_m(l)dP(l) = \frac{RT}{P(g)}dP(g) \Rightarrow \frac{V_m(l)dP(l)}{RT} = \frac{dP(g)}{P(g)} \tag{1.11}$$

In the absence of any additional external pressure, the pressure experienced by the drop (inner) is equal to its vapor pressure $P_1(g)$, i.e. vapor pressure of the planar surface. In the presence of an additional pressure ΔP, the pressure on the drop (inner) is $P_1(g) + \Delta P$ while that on the outer vapor phase (i.e. vapor pressure of the drop) is $P_2(g)$. This is because the vapor pressure of a liquid depends on the pressure applied to the liquid. Integrating Equation (1.11) along these limits, as in (1.12), gives (1.13)

$$\frac{V_m(l)}{RT}\int_{P_1(g)}^{P_1(g)+\Delta P} dP(l) = \int_{P_1(g)}^{P_2(g)} \frac{dP(g)}{P(g)} \tag{1.12}$$

$$\frac{V_m(l)\Delta P}{RT} = \ln\left(\frac{P_2(g)}{P_1(g)}\right) \tag{1.13}$$

The ΔP also represents the pressure difference across the drop interface and is given by the Laplace Equation (1.5). Therefore, substituting Equation (1.5) into (1.13) gives the Kelvin Equation (1.14)

$$\ln\left(\frac{P_2(g)}{P_1(g)}\right) = \frac{2\gamma_{la}M}{r\rho RT}\left(\text{Noting that } V_m(l) = \frac{\text{molar mass } M}{\text{density } \rho}\right) \tag{1.14}$$

For a cavity in a liquid, the external pressure $P_1(g)$ of the liquid is greater than the internal vapor pressure $P_2(g)$, and its Kelvin equation is

$$\ln\left(\frac{P_2(g)}{P_1(g)}\right) = -\frac{2\gamma_{la}M}{r\rho RT} \tag{1.15}$$

Example 1.2: Calculating the Vapor Pressure of Liquid Drops

For water of surface tension ~72 mN m^{-1}, vapor pressure 23.8 Torr and density 1000 kg m^{-3}, all at 25°C, calculate the vapor pressure of a spherical water drop whose diameter is 20 nm.

Method

Apply the Kelvin equation after converting the surface tension to N m^{-1}, temperature to Kelvin and diameter to metre and note that the molar mass of water is ~18 × 10^{-3} kg mol^{-1}.

Answer

$$\ln\left(\frac{P_2(g)}{23.8 \text{ Torr}}\right) = \frac{2 \times 2 \times 72 \times 10^{-3} \text{ N m}^{-1} \times 18 \times 10^{-3} \text{ kg mol}^{-1}}{20 \times 10^{-9} \text{ m} \times 1000 \text{ kg m}^{-3} \times 8.314 \text{ J K}^{-1} \text{mol}^{-1} \times 298 \text{ K}}$$

$$P_2(g) = 23.8 \text{ Torr} \times \exp(0.1046) = 26.4 \text{ Torr}$$

Notice that the vapor pressure of the drop is greater than that of the liquid (*i.e.* a flat surface).

1.3 MEASUREMENT OF SURFACE TENSION

There are many methods (classical and modern) for measuring the surface tension of fluid interfaces, and the choice of method depends on a given system. These methods are categorized into five groups.

- **Group I** The surface (or interfacial) tension is measured directly using a microbalance. Examples include the Wilhelmy plate and du Noüy ring methods.
- **Group II** The value for surface tension is obtained (indirectly) from the measurement of capillary pressure. Examples are the maximum bubble pressure and growing drop methods.

- **Groups III and IV** Surface tension values are obtained from the analysis of the equilibrium between capillary and gravity forces. Group III relies on the balance between surface tension forces and a variable volume of liquid. Examples of typical methods include the capillary rise and the drop volume methods. In the case of group IV, the degree of distortion of a fixed volume of liquid under the influence of gravity is measured. Typical examples include the pendant and sessile drop methods.
- **Group V** Surface tension values are obtained from the degree of distortion of a fixed volume of liquid by centrifugal forces. Examples include the spinning drop and micropipette methods. These techniques are used for the measurement of ultralow interfacial tensions.

1.3.1 DIRECT MEASUREMENT USING A MICROBALANCE

This method is very suitable for measuring the liquid(l)-air(a) γ_{la} interfacial tension, also known as liquid surface tension. The direct measurement of γ_{la} using a microbalance requires a plate, ring, rod, or other probes of simple shape and involves bringing them in contact with the interface. Because the liquid completely wets the probe, it adheres to the probe and climbs as a result of capillary forces, thereby increasing the interfacial area and leading to a force that tends to pull the probe towards the plane of the interface, as illustrated in Figure 1.4. The restoring force is directly related to the

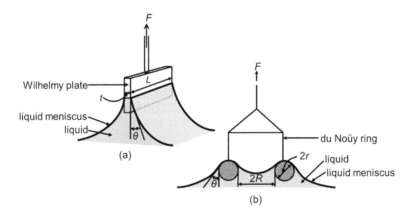

FIGURE 1.4 Schematic description of the (a) Wilhelmy plate and (b) du Noüy ring methods of surface tension measurement.

interfacial tension and is measured with a microbalance. The force F acting along the three-phase contact line represents the weight of the liquid meniscus standing above the plane of the liquid-air interface. The F is measured by the microbalance and used for the calculation of the interfacial tension using Equation (1.16), where X

stands for the parameter $2(L + t)$ of the three-phase contact line while θ stands for the contact angle between the liquid-liquid meniscus in contact with the surfaces of the probe.

$$\gamma_{la} = \frac{F}{X \cos \theta} \tag{1.16}$$

Liquid-liquid, *e.g.* oil-water, interfacial tensions have also been measured using this method, especially when the tension is above 20 mN m⁻¹ (Binks *et al.* 2010). Here, the force required to break the interface or detach the probe from the interface is measured and used for the calculation.

1.3.1.1 The Wilhelmy Plate Method

A vertical thin plate, as described in Figure 1.4a, is used. The plate is made of roughened platinum-iridium alloy or platinum. The plate is cleaned from organic contaminants by an organic solvent and then flamed with a Bunsen flame. Both roughening and cleaning of the plate surfaces help the test liquid to wet it perfectly, thus making $\theta = 0°$. Glass, mica, and steel can be used in place of the alloy (Rusanov and Prokhorov 1996). This is particularly the case when the interfacial tension between heavy nonpolar liquids such as carbon tetrachloride and an immiscible, but lighter, polar liquid like water is to be measured. This usually requires the plate to be hydrophobic. This is achieved by using fluorinated polymers, which are inherently hydrophobic. In some cases, the surfaces of the alloy are made hydrophobic by coating them with organic amines and used for the experiment.

The Wilhelmy plate experiment is performed in both static and detachment modes. In the static mode, the plate remains in contact with the liquid during the entire cycle of the experiment. The plate is held stationary relative to the liquid interface and the force F acting vertically on the plate by the liquid meniscus when the liquid makes contact with the plate is measured using a microbalance. The force on the plate is equal to the weight of the liquid meniscus uplifted over the horizontal surface. In the detachment mode, the force F required to detach the plate from the interface is measured. In both cases, the F is then used to calculate the corresponding interfacial tension value by using Equation (1.16). Modern instruments use plates of standard dimensions and weight so that one does not have to be bothered by their precise values during calculation of the interfacial tension. Adsorption of organic compounds from the laboratory environment or test solution drastically affects the experimental values and care should be taken to avoid it (Rusanov and Prokhorov 1996).

1.3.1.2 The du Noüy Ring Method

This is a detachment technique where the maximum force F required to pull a wire ring off an interface is measured by a microbalance, Figure 1.4b, and used for calculation of interfacial tension. The ring (radius $R = 2$–3 cm) is usually made of platinum or platinum-iridium alloy just like the Wilhelmy plate. The radius r of the wire ranges from 1/30 to 1/60 of that of the ring (Vold and Vold 1983). Because of the additional volume of liquid lifted during the detachment of the ring from the interface, Equation (1.16) is corrected to (1.17) before the interfacial tension calculation.

$$\gamma_{la}(corr) = \frac{\overbrace{F}^{\gamma_{la}}}{X\cos\theta} \times f, \text{ where } X = 4\pi R \text{ and } \theta = 0° (\text{perfect wetting}) \quad (1.17)$$

and

$$f = \begin{cases} 0.725 + \sqrt{\dfrac{3.63\times10^{-3}\gamma_{la}}{\pi^2\Delta\rho R^2} - \dfrac{1.679r}{R} + 0.04534} & \gamma_{la} < 25 \text{ mN m}^{-1} \\[3mm] \dfrac{\Delta\rho g R^2}{4\pi\gamma_{la}} & \gamma_{la} > 25 \text{ mN m}^{-1} \end{cases} \quad (1.18)$$

The f is a correction factor, whose value ranges from 0.75 to 1.05, depending on the dimensions (R and r) of the ring. The $\Delta\rho$ is the difference in liquid-air or liquid-liquid density. The values of f, in relation to R/r (for $\theta = 0$), have been reported by Rusanov and Prokhorov (1996) and can also be calculated using Equation (1.18). The $\gamma_{la}(corr)$ represents the corrected surface tension. The γ_{la} from modern computerized instruments does not require any correction as f is already incorporated in the software and can be taken as the interfacial tension of the system.

For high accuracy in the measurement of γ_{la}, the plane of the ring is required to be parallel to the interface. Deformation of the ring (a very fragile probe) during handling and cleaning drastically affects measurements. It is also important that the necessary liquid wets the ring perfectly ($\theta = 0°$) so that the measurement will be possible and additional correction will not also be needed. The method is more accurate than other detachment methods, provided the necessary precautions are observed.

Example 1.3: Determination of Surface Tension Using the du Nöuy Ring Method

In an experiment to measure the surface tension of a liquid at 20°C using a du Nöuy ring of radius 9.55 mm made of wire of radius 0.18 mm, the maximum force required to detach the ring from the liquid-air interface was ~9 mN. Calculate and correct the corresponding value of the surface tension if the density of the liquid and air at the prevailing condition is 0.998 and 0.0098 g cm^{-3}, respectively.

Method

Convert the radius of the ring and that of the wire to metre and the density of the liquid and air phases to kg m^{-3} and the detachment force to N. Assume that the liquid wets the ring perfectly (i.e. $\theta = 0°$). Use Equation (1.16) to calculate the surface tension that corresponds to the detachment force and then use the appropriate f in Equation (1.18) to correct the surface tension value.

Answer

$$\gamma_{la} = \frac{F}{4\pi R} = \frac{9 \times 10^{-3}\ N}{4\pi \times 9.55 \times 10^{-3}\ m} = 0.075\ N\ m^{-1} = 75\ mN\ m^{-1}$$

Because $\gamma_{la} > 25\ mN\ m^{-1}$,

$$f = \frac{\Delta \rho g R^2}{4\pi \gamma_{la}} = \frac{(998 - 9.8)\ kg\ m^{-3} \times 9.8\ m\ s^{-2} \times (9.55 \times 10^{-3})^2\ m^2}{4\pi \times 75 \times 10^{-3}\ N\ m^{-1}} = 0.937$$

$$\gamma_{la}(corr) = \gamma_{la} \times f = 75\ mN\ m^{-1} \times 0.937 = 70\ mN\ m^{-1}$$

1.3.2 MEASUREMENT OF CAPILLARY PRESSURE: THE MAXIMUM BUBBLE PRESSURE METHOD

It has been stated earlier that there is a pressure difference ΔP between fluids on either side of a curved interface with the concave side having a higher pressure, and that surface tension is the reason why liquid drops and air bubbles are spherical. The pressure difference can be calculated using the Young-Laplace equation if the interfacial tension and the radii of curvature are known. When the interfacial tension is desired, the pressure difference is measured by using a pressure sensor or by observing a capillary rise and then used for interfacial tension calculation provided the radii of curvature are known (Young 1855).

The maximum bubble pressure method has been used to measure interfacial tension for many years and has now been modified to measure even dynamic (*i.e.* non-equilibrated) interfacial tension. In the maximum bubble pressure method (Figure 1.5), the maximum pressure P_{max} needed to force a gas bubble out of a capillary

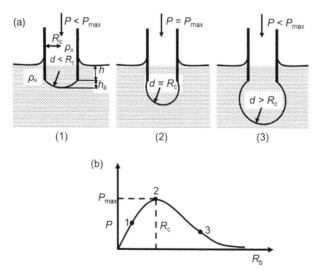

FIGURE 1.5 Schematic description of the maximum bubble pressure method of surface tension measurement: (a) an illustration of the shape of the bubble at three different stages of growth and (b) a plot of the inner pressure P of the bubble against its radius R_b.

tube of radius R_c into a liquid is measured. The P_{max} is equal to the sum of the capillary pressure ΔP caused by the interfacial tension and the hydrostatic pressure (ρgh) caused by the liquid column above the orifice of the capillary as shown in Equation (1.19).

$$P_{max} = \Delta P + \rho gh \tag{1.19}$$

The ΔP is related to the height h_i of an imaginary column of liquid through Equation (1.20), in which the difference between the liquid ρ_A and air ρ_B density is $\Delta \rho = \rho_A - \rho_B$.

$$h_i = \frac{\Delta P}{\rho gh} \tag{1.20}$$

Sugden (1922) has derived a formula, Equation (1.21), that connects h_i with the Laplace capillary constant $K = 2\gamma_{la}/\Delta \rho g$ and the bubble meniscus. The Y is K^2/h_i, β is $2d^2/K^2$, h_b is the height of the bubble while R_b is the radius of the bubble. Sugden (1922) further showed that R_c/K values are dependent on a given value of R_c/K within the range of $0 < R_c/K \leq 1.5$ and used them to calculate γ_{la} following a set of procedures.

$$\frac{R_c}{Y} = \frac{R_c}{R_b} + \left(\frac{R_c}{K}\right)\left(\frac{h_b}{R_b}\right)\left(\frac{\beta}{2}\right)^{0.5} \tag{1.21}$$

Values of γ_{la} can also be calculated using Equation (1.22), but such values are less accurate.

$$\gamma_{la} = \frac{\Delta P R_c}{2}\left(1 - \frac{2R_c\Delta \rho g}{3\Delta P} - \frac{1}{6}\left(\frac{R_c\Delta \rho g}{\Delta P}\right)^2\right) \tag{1.22}$$

1.3.3 MEASUREMENT BASED ON THE BALANCE BETWEEN CAPILLARY AND GRAVITY FORCES

1.3.3.1 The Capillary Rise Method

It was stated in Section 1.2.2 that when a capillary tube is inserted vertically upwards in a liquid, the liquid may either rise in the tube above the liquid-air interface or fall in the tube below the liquid-air interface, with the degree of rise or fall depending on the tube's inner diameter. (The rise is higher in tubes of relatively small inner diameter while the fall is lower in tubes of relatively small inner diameter.) The adhesive and cohesive forces are responsible for this phenomenon. The adhesive force causes the rise or fall while the cohesive (due to surface tension) force causes all the liquid molecules to follow the upward or downward pull (Ehlers and Goss 2003). In the capillary rise method, the height h (above the liquid-air interface) of the liquid meniscus in a tube of inner radius r, due to capillarity, is measured as illustrated in Figure 1.6. The rise of liquid in

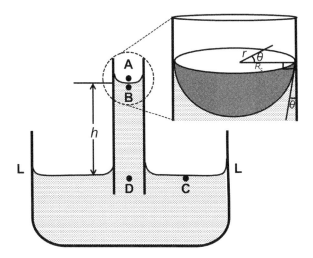

FIGURE 1.6 Schematic illustration of the capillary rise method of surface tension measurement, showing the rise height h and a meniscus of radius R_c making a contact angle of θ with the tube of inner radius r.

capillaries is best treated in terms of the pressure difference across the meniscus (McCaughan 1987, Tabor 1991).

In the absence of air and the vapor pressure of the liquid, the pressure at points A and C are equal to zero (*i.e.* $P_A = P_C = 0$), and by Pascal's principle, the pressures at points C and D are also equal (*i.e.* $P_C = P_D = 0$). In the presence of air and liquid vapor pressure, the pressure at point B is less than that at point A while the pressure at point D is greater than that at point B. This gives rise to a pressure gradient which produces a force that acts vertically upwards over the entire bulk of the liquid column in opposition to the weight of the straight-wall column. The difference in pressure across the curved interface ($P_A - P_B = \Delta P$) is balanced by the weight of the column in the gravitational field, ρgh, where ρ is the density of the liquid, g is the acceleration due to gravity, and h is the rise height. It is important to note that the difference in pressure at points D and B, $P_D - P_B$, also equals ΔP. Based on Equation (1.5),

$$\Delta P = \frac{2\gamma_{la}}{R_c} = \rho gh \tag{1.23}$$

Based on Figure 1.6, $R_c = r/\cos\theta$ and Equation (1.23) becomes (1.24), which gives the capillary rise.

$$h = \frac{2\gamma_{la}\cos\theta}{\rho gr} \Rightarrow \gamma_{la} = \frac{\rho grh}{2\cos\theta} \tag{1.24}$$

The θ is $0°$ for the case of complete wetting and $180°$ for the case of non-wetting. Equation (1.24), however, does not take cognizant of the liquid that stands higher

than point B in the curved portion of the meniscus thereby affecting the calculation of γ_{la}. Bikerman (1970) has considered this for a more precise calculation of
γ_{la}. Supposed that the volume of liquid lifted (above LL) is V, the rise height h is V
divided by the cross-sectional area (πr^2) of the capillary. The h is also the vertical
distance between LL and bottom of the meniscus. Because the h in the latter and the
former differ by the factor $r/3$, it is added to the rise height (Bikerman 1970) so that

$$\gamma_{la} = \frac{\rho g r}{2\cos\theta}\left(h + \frac{r}{3}\right) \tag{1.25}$$

Measuring h for a liquid of known density ρ in a tube of inner radius r where it is
supposed that θ is $0°$ allows the precise calculation of the surface tension γ_{la} using
Equation (1.25). However, Equation (1.25) is limited to tubes of relatively small
diameters (where $r \ll h$), where the meniscus is approximately spherical. For relatively large tubes where the meniscus is non-spherical, Equation (1.26) is used.

$$\gamma_{la} = \frac{\rho g r}{2\cos\theta}\left[h + \frac{r}{3} - \frac{0.1288r^2}{h} + \frac{0.1312r^3}{h^2}\right] \tag{1.26}$$

Equation (1.24) can also be obtained by considering the surface forces around the
periphery of the meniscus (Tabor 1991). The periphery length of the meniscus is $2\pi r$
and for a contact angle of θ, the upward force due to surface tension is $2\pi r\gamma_{la}\cos\theta$.
In the case of straight-wall capillaries, this is balanced by the downward weight
($mg = \rho Vg = \pi r^2 h\rho g$) of the liquid column. The m is mass of the liquid column and
V is its volume. At equilibrium, $2\pi r\gamma_{la}\cos\theta = \pi r^2 h\rho g$, as in Equation (1.24). This
derivation breaks down when the tube does not have a uniform cross section.

The capillary rise method measures γ_{la} accurately if technical problems associated with it (*e.g.* fabrication of a uniform bare capillary tube and precise determination of its inner diameter) are overcome. This method, nonetheless, does not measure
the interfacial tension between two liquids.

Example 1.4: Calculating Surface Tension from Capillary Rise Experiment

Distilled water at 40°C (density 992.2 kg m^{-3}) rose to a height of 2.8 mm in a capillary tube of inner diameter 7.62 mm. Calculate the surface tension of the water at
this temperature if it wets the tube perfectly.

Method

Convert the rise height h and inner diameter to metre and divide the diameter
by 2 to obtain the radius r. Use Equation (1.26) to calculate the surface tension
since r is not far-far less than h, noting that $\cos\theta = 1$ for perfect wetting and that
$g = 9.8$ m s^{-2}.

Answer

$$\frac{\rho g r}{2 \cos \theta} = \frac{992.2 \, \text{kg m}^{-3} \times 9.8 \, \text{m s}^{-2} \times 3.81 \times 10^{-3} \text{m}}{2 \times \cos \theta} = 18.5 \, \text{N m}^{-2}$$

$$h + \frac{r}{3} - \frac{0.1288 r^2}{h} + \frac{0.1312 r^3}{h^2}$$

$$= 2.8 \times 10^{-3} \, \text{m} + \frac{3.81 \times 10^{-3} \, \text{m}}{3} - \frac{0.1288 (3.81 \times 10^{-3})^2 \, \text{m}^2}{2.8 \times 10^{-3} \, \text{m}}$$

$$+ \frac{0.1312 (3.81 \times 10^{-3})^3 \, \text{m}^3}{(2.8 \times 10^{-3})^2 \, \text{m}^2} = 4.3 \times 10^{-3} \, \text{m}$$

Therefore,

$$\gamma_{\text{la}} = (18.5 \, \text{N m}^{-2})(4.3 \times 10^{-3} \, \text{m}) = 80.2 \, \text{mN m}^{-1}$$

Example 1.5: Walking on Water

Using surface tension for support, many insects are able to walk on the surface of water. Consider a six-legged water strider (Figure 1.7) walking on water. A depression, of contact angle θ, is formed around the water when each foot steps on the water. The surface tension of the water produces an upward force on the water that tends to restore the water surface to its normal shape. If the surface tension and the density of the water are 65 mN m^{-1} and 995 kg m^{-3} respectively and the mass M, length L, and radius r of the tarsal segment of the insect are 2×10^{-5} kg, 0.15 cm and 0.006 cm respectively, find the angle of depression θ on each foot.

Method

The resultant vertical component of the surface tension force F on each foot is $2\gamma_{\text{la}} \sin \theta \times P$. The P ($= 12L$) is the combined contact perimeter of the tarsal segment (provided all of them are equal). This force bears the weight (Mg, where $g = 9.8$ m s^{-2}) of the insect. At equilibrium,

$$2\gamma_{\text{la}} \sin \theta \times P = Mg$$

Answer

$$\Rightarrow \theta = \sin^{-1} \left(\frac{Mg}{2\gamma_{\text{la}} \times P} \right)$$

$$\theta = \sin^{-1} \left(\frac{2 \times 10^{-5} \, \text{kg} \times 9.8 \, \text{m s}^{-2}}{2 \times 65 \times 10^{-3} \, \text{N m}^{-3} \times 12 \times 1.5 \times 10^{-3} \, \text{m}} \right) = 4.8°$$

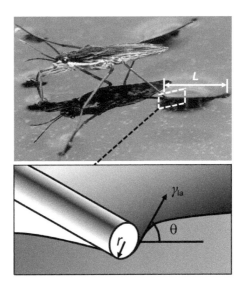

FIGURE 1.7 The legs of water striders are covered with hair and made effectively non-wetting. When on water, the tarsal segment (radius *r*) of its legs rest on the free surface and as a result, the free surface makes an angle θ with the horizontal. The weight is supported by a combination of buoyance and curvature forces. The surface tension γ_{la} of the water produces an upward force on the water that tends to restore the water surface to its normal shape. (Courtesy of BBC Nature courtesy of FLPA/Alamy Stock Photo; sketch redrawn from Hu, D.L. *et al., Nature*, 424, 663–666, 2003.)

1.3.3.2   The Drop Volume or Weight Method

For the drop volume or weight method, the volume *V* or weight of a liquid drop falling from a capillary tube of radius *r* is measured. The weight *w* of the liquid drop falling off the capillary is related to the interfacial tension according to Equation (1.27).

$$W = \Delta\rho gV = 2\pi\gamma_{la}f'\left(\frac{r}{V^{1/3}}\right) \qquad (1.27)$$

The Δρ is the density difference between the liquid phase and air and *f′* is a correction factor required to account for the portion of the drop that is not released from the capillary during detachment. The *f′* is a function of $r/V^{1/3}$. The values of *f′* have been reported by Harkins and Brown (1919) and can also be calculated by using Equation (1.28).

$$f'\left(\frac{r}{V^{1/3}}\right) = 0.167 + 0.193\left(\frac{r}{V^{1/3}}\right) - 0.0489\left(\frac{r}{V^{1/3}}\right)^2 - 0.0496\left(\frac{r}{V^{1/3}}\right)^3 \quad (1.28)$$

In practice, many liquid drops are collected so as to have an accurate measurement of the weight or volume. This is because the individual drops themselves are small.

In modern instruments, the volume of the liquid and the number of drops released from the capillary are determined very accurately, therefore making it easier to calculate the weight or volume of the individual liquid drops. The capillary tubes used for the technique are typically made of glass or metal (occasionally). Glass is commonly used because it is transparent, wetted by many liquids, and relatively easy to clean compare to a metal. Measurements with this technique are very simple, but are sensitive to vibrations which cause premature separation of the liquid drop from the end of the capillary before the drop reaches the critical size (*i.e.* the minimum size required for separation). Also, in multicomponent solutions involving adsorption of surface-active solutes, the measurements hardly reflect equilibrium saturation of the solutes at the interface.

1.3.4 ANALYSIS OF GRAVITY DISTORTED DROPS

Surface tension causes interfaces to behave like elastic membranes that tend to compress the liquid. In the absence of gravity, a liquid surface has an inherent ability to be spherical. This minimizes the interfacial area per unit volume of liquid and consequently the excess energy of the surface. Two types of liquid drops (the pendant and the sessile drops) are known. A pendant drop is simply a suspended (or hanging) drop formed at the end of a thin tube and held by surface tension forces, while a sessile drop is simply a liquid drop resting on a substrate. The shape of pendant drops depends on the interplay of gravitational, capillary, and surface tension forces, while that of sessile drops depends only the interplay of gravitational and surface forces. Relatively small sessile drops, where surface tension forces dominate over gravitational forces, are spherical. However, relatively large ones, where gravitational forces dominate over surface tension forces are puddle-shaped. The analysis of the shape of a pendant or sessile drop allows the calculation of the surface tension value of the given liquid.

1.3.4.1 The Pendant Drop Method

Consider the shape of a liquid drop suspended vertically at the tip of a thin tube of inner radius r (Figure 1.8). Its shape is a consequence of the balance between the capillary, gravitational, and surface tension forces, and it is said to be in hydromechanical equilibrium. The drop is rotationally symmetrical in the Z-axis direction (*i.e.* axisymmetrical). If the tangent at the intersection of the Z-axis with the apex of the drop forms the X-axis, the drop profile can be given by pairs of x, z values in the X-Z plane. From the Young-Laplace equation, the pressure difference ΔP across a curved interface (liquid-air, in this case) of interfacial tension γ_{la}, and principal radii of curvature R_1 and R_2 can be written as given in Equation (1.29). The analogues equation (*i.e.* the ΔP), with respect to a point $Q(x, z)$, for a pendant drop formed in air saturated with its vapor, whose shape is controlled mainly by gravity and surface tension forces, is Equation (1.30). The R_1 and R_2 refer to the principal radii of curvature at the point $Q(x, z)$. R_1 is the principal radius of curvature in the drop cross-section that includes the Z axis, while R_2 is that in the plane perpendicular to the former. ϕ is the inclination angle made by the tangent at the point $Q(x, z)$ and the X coordinate axis. Because the drop is axisymmetric with respect to the vertical axis, the pressure difference ΔP_O in the apex of the drop can be calculated from Equation (1.31), in which b is the principal radius of curvature at the origin O. Once in hydromechanical equilibrium,

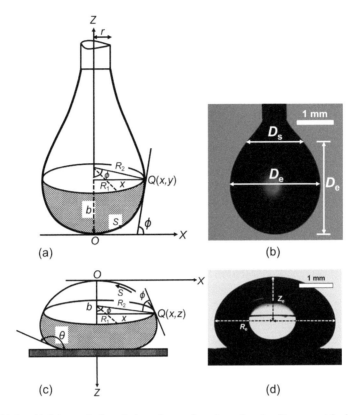

FIGURE 1.8 (a) Schematic description of a pendant drop showing its geometric description. (b) A water pendant drop showing the diameter D_e at the equatorial plane and the diameter D_s at a distance D_e from the tip of the drop. (c) Schematic description of a sessile drop showing its geometric description. (d) A glycerol sessile drop (10 µL) showing the equatorial radius R_e and the equatorial height Z_e.

ΔP and ΔP_O are related to the hydrostatic pressure $\Delta \rho g z$ as given in Equation (1.32). Where $\Delta \rho$ is the density difference between the drop and the vapor phase, g is the acceleration due to gravity and z is the height above the drop apex. The substitution of Equations (1.30) and (1.31) into (1.32) gives (1.33). Equation (1.33) can also be written as in (1.34), where β is the shape parameter. Because of the presence of symmetry in the drop, the mathematical description of the interface can be reduced to the description of only the meridian section (hatched area in Figure 1.8a). The best representation of the meridian curve is the parametric form given in Equation (1.35), with S being the arc-length measured from the origin O. The X and Z in the equation are single-valued function of S.

$$\Delta P = \gamma_{la} \left(\frac{1}{R_1} + \frac{1}{R_2} \right) \tag{1.29}$$

$$\Delta P = \gamma_{la} \left(\frac{1}{R_1} + \frac{\sin \phi}{X} \right) \tag{1.30}$$

$$\Delta P_O = \frac{2\gamma_{la}}{b} \tag{1.31}$$

$$\Delta P = \Delta P_O - \Delta \rho g z \tag{1.32}$$

$$\gamma_{la}\left(\frac{1}{R_1} + \frac{\sin\phi}{X}\right) = \frac{2\gamma_{la}}{b} - \Delta \rho g z \tag{1.33}$$

$$\frac{1}{R_1/b} + \frac{\sin\phi}{X/b} = 2 + \beta\left(\frac{z}{b}\right) \text{where } \beta = -\frac{\Delta\rho g b^2}{\gamma_{la}} \tag{1.34}$$

$$X = X(S) \text{ and } Z = Z(S) \tag{1.35}$$

A geometric analysis of the meridian section gives

$$\frac{1}{R_1} = \frac{d\phi}{dS} \tag{1.36}$$

$$\frac{dX}{dS} = \cos\phi \tag{1.37}$$

and

$$\frac{dZ}{dS} = \sin\phi \tag{1.38}$$

(Equation [1.36] shows that R_1 is the rate of change of the turning or inclination angle ϕ with respect to the arc-length parameter S.)

The differential form of Equation (1.34) is obtained by substituting Equation (1.36) into (1.34) followed by rearrangement. This gives Equation (1.39).

$$\frac{d\phi}{dS} = 2 + \beta z - \frac{\sin\phi}{X} \tag{1.39}$$

where (from differential geometry),

$$\frac{d\phi}{dS} = \left(\frac{d^2Z}{dX^2}\right)\left(1 + \left(\frac{dZ}{dX}\right)^2\right)^{-\frac{3}{2}} \tag{1.40}$$

and

$$\frac{\sin\phi}{X} = \left(\frac{dZ}{dX}\right)\left(\frac{1}{X}\right)\left(1 + \left(\frac{dZ}{dX}\right)^2\right)^{-\frac{1}{2}} \tag{1.41}$$

Equations (1.37 through 1.39) and the boundary conditions given in Equation (1.42) represent a set of first-order differential equations for X, Z, and ϕ as functions for S.

$$X(0) = Z(0) = \phi(0) \qquad (1.42)$$

It is difficult to solve the differential equation in (1.39) directly. Bashforth and Adams (1883) have solved this equation using numerical integration techniques and generated tables for the drop contours. Surface tension values can be obtained using these tables by fitting the experimental drop contour to the theoretical curve but this procedure is very cumbersome. As a result, the method has been simplified by Andreas *et al.* (1937). The Andreas *et al.* (1937) method is based on measuring the drop diameter D_e at the equatorial plane and the diameter D_s at a distance D_e from the tip of the drop, as illustrated in Figure 1.8b, and solving the fundamental equation. The calculation of surface tension values from the experimental data requires the use of a correction factor $1/H$, a function of the ratio D_s/D_e denoted by S_F, in which S_F is the shape factor. The surface tension values are obtained from the following empirical formula:

$$\gamma_{la} = \frac{\Delta \rho g D_e^2}{H} \qquad (1.43)$$

Values for $1/H$ *versus* S_F have been reported by many authors (Andreas *et al.* 1937, Fordham 1948, Niederhauser and Bartell 1948–1949, Stauffer 1965, and Roe *et al.* 1967, Bashforth and Adams 1883) with varying degrees of accuracy. Values of $1/H$ can also be calculated from Equation (1.44). The $B_i (i = 0, 1, 2, 3, 4)$ and a are empirical constants for a given range of S_F as given in Table 1.2.

$$\frac{1}{H} = \frac{B_4}{S_F^a} + B_3 S_F^3 - B_2 S_F^2 + B_1 S_F - B_0 \qquad (1.44)$$

With current progress in image analysis and data acquisition, it is possible to obtain an image of a drop directly using the video frame grabber of a digital camera. This also allows calculation of the surface tension value of the drop. The surface tension value of the drop is obtained from the analysis of its profile using specialized

---

TABLE 1.2

Range of S_F and Values of the Corresponding Constants

Range of S_F	a	B_4	B_3	B_2	B_1	B_0
0.401–0.46	2.56651	0.32720	0	0.97553	0.84059	0.18069
0.46–0.59	2.59725	0.31968	0	0.46898	0.50059	0.13261
0.59–0.68	2.62435	0.31522	0	0.11714	0.15756	0.05285
0.68–0.90	2.64267	0.31345	0	0.09155	0.14701	0.05877
0.90–1.00	2.84636	0.30715	−0.69116	−1.08315	−0.18341	0.20970

software, built-in house, in many cases. This, generally, requires solving the fundamental equation. In practice, the drop is imaged and the γ_{la} is considered as a fitting parameter. The value of the γ_{la} is adjusted until the solution of Equation (1.39) agrees with the experimental result obtained from the drop imaging.

For the purpose of good quality and reproducible results, the needle used for hanging the liquid drop is required to be clean, and climbing of the interface over its outer surface should also be avoided. Needles made of stainless steel or glass that is relatively easy to clean with acids, bases, and organic solvents should be used. More importantly, the needle's diameter should be less than $0.5D_e$ but should not be too small as this reduces the value of D_s and hence the accuracy of the surface tension value obtained.

1.3.4.2   Sessile Drop Method

This method involves the analysis of the profile of a liquid drop resting on a solid substrate, as illustrated in Figure 1.8c. The substrate is required to be poorly wetted by the liquid drop, *i.e.* $\theta > 90°$. The analogues of Equation (1.34) for a sessile drop is:

$$\frac{1}{R_1/b} + \frac{\sin\phi}{X/b} = 2 + \beta\left(\frac{z}{b}\right) \text{ where } \beta = \frac{\Delta\rho g b^2}{\gamma_{la}} \tag{1.45}$$

Equation (1.45) cannot be integrated (or solved) in finite terms, but Bashforth and Adams (1883) have given the numerical solution for surfaces of revolution and provided values for X/b, z/b, and V/b^3, with V being the volume between the origin and a horizontal plane at the level z, for given values of ϕ and β. However, in practice, only the equatorial height Z_e and the equatorial radius R_e, Figure 1.8d, are measurable and for a large drop ($R_e > 2$ cm, *e.g.* water),

$$\frac{\alpha^2}{R_e^2} = \frac{Z_e^2}{R_e^2} \text{ where } \alpha = \sqrt{\frac{2\gamma_{la}}{\Delta\rho g}} \tag{1.46}$$

Thus, α and hence the surface tension value can be found at once. Table of values for α^2/R_e^2 *versus* Z_e^2/R_e^2 are available in the literature (Taylor and Alexander 1944). There are also α^2/R_e^2 *versus* Z_e^2/R_e^2 values for smaller ($R_e \leq 2.2$ cm, *e.g.* water) and approximately flat drops based on Equation (1.46).

The pendant and sessile drops methods are very attractive as they do not require advanced instrumentation. The experimental set-up requires a camera with a low-magnification lens to record the shape of the drop. One drawback of the sessile drop method is the difficulty associated with locating the equator precisely and measurement of Z_e.

The pendant and sessile drop methods can also be used for liquid-liquid interfacial tension measurement. To do this, the profile of the drop (of the denser liquid phase) is obtained in the presence of the less dense one which forms the continuous phase. The drop image is analyzed, similarly, taking $\Delta\rho$ as the density difference between the two liquid phases.

1.3.4.3 Spinning Drop Tensiometry and Other Methods

Measurement of interfacial tension is also possible from spinning drop tensiometry, microtensiometry, micropipette, and atomic force microscopy, all illustrated in Figure 1.9. A spinning drop tensiometer is capable of measuring interfacial tension values lower than 10^{-4} mN m^{-1} (Sottmann and Strey 1997). Spinning drop tensiometry is based on the fact that gravity has little or no effect on the shape of a fluid drop (less dense, 1) suspended in the bulk of another fluid (more dense, 2) contained in a horizontal capillary tube rotating about its horizontal axis where the two fluids remain completely immiscible, Figure 1.9a. The fluid drop is ellipsoidal at relatively low rotational speed ω, but elongates and becomes cylindrical at relatively high ω where centrifugal forces dominate. Under the latter condition, the radius R_c of the cylindrical fluid is detected by the fluid interfacial tension γ_{12}, the density difference $\Delta\rho$ between the fluid phases and ω. Using this as a basis, the interfacial tension is calculated quite accurately from Equation (1.47) provided the ratio of the length L_c, excluding the hemispherical ends, of the cylindrical fluid to its diameter D_c is ≥ 4 (Vonnegut 1942). In practice, Equation (1.47) is modified prior to the calculation. The modification takes cognizance of the refractive indexes of the bulk fluid and the

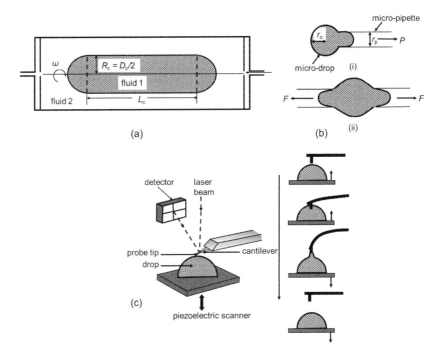

(a)

(b)

(c)

FIGURE 1.9 Schematic illustration of (a) spinning drop tensiometry, (b) micropipette tensiometry, showing the (i) sucking of a microdrop into a micropipette by a pressure P and (ii) deformation of a microdrop from the two-pipette method by the two opposing forces F, and (c) atomic force microscopy and its application in the measurement of interfacial tension for microdrops.

thermostating oil, which maintains the temperature of the tensiometer during the measurement, and the magnification of the instrument lens used.

$$\gamma_{12} = \frac{\Delta\rho\omega^2 R_c^3}{4}\left(1 + \frac{2R_c}{3L_c}\right) \tag{1.47}$$

Example 1.6: Determining Interfacial Tension Values from Spinning Drop Tensiometry

In order to measure the interfacial tension (at 25°C) between sunflower oil (1, $\rho \approx 0.92$ g cm^{-3}) and 20 cS polydimethylsiloxane (2, $\rho \approx 0.95$ g cm^{-3}), a drop of sunflower oil (less dense) was spun in the bulk of the polydimethylsiloxane (denser phase) at a rotation speed N of 8268 revolutions per min (cf. ω/radians per second) to obtain a cylindrical sunflower oil drop. The D_c and L_c of the sunflower oil drop were measured to be 3.7 and 18 microscope graticule, respectively, using a lens of magnification × 2.5 and instrument [RED (site 04)]. If the refractive indexes (at 25°C) of the thermostating oil (n_T) and polydimethylsiloxane (n_2) are 1.41 and 1.40, respectively, what is the sunflower oil-polydimethylsiloxane interfacial tension?

Method

For the spinning drop tensiometry, the sunflower oil (1)-polydimethylsiloxane (2) interfacial tension γ_{12} is given by Equation (1.47). However, Equation (1.47) would be modified to allow for the conversion of measured values of N into ω and of D_c into R_c.

Answer

$$\omega\,(\text{radians per second}) = \left(\frac{2\pi}{60}\right)N = (0.10472)\,N \tag{1.48}$$

$$R_c = xR_c\left(\frac{n_T}{n_2}\right) \tag{1.49}$$

The factor x in Equation (1.49) is used to convert measured D_c (in graticule units) to R_c (in metres) and it includes both the magnification of the microscope and the factor 2 (for conversion of R_c to D_c). In addition to the magnification of the microscope, there is additional magnification caused by the lens effect of the curved capillary which depends on the ratio of the n_T to n_2 (Coucoulas et al. 1983). The substitution of Equations (1.48) and (1.49) into (1.47) gives (1.50), which can be used at once to obtain the interfacial tension.

$$\gamma_{12} = N^2 D_c^3 A\Delta\rho\left(\frac{n_T}{n_2}\right)^3\left[1 + \left(\frac{n_T D_c}{n_2}\right)\bigg/3L_c\right] \tag{1.50}$$

The constant A, given by Equation (1.51), contains all sample invariant conversion factors. The factor of 10^6 arises from the conversion of density unit from g cm^{-3} to kg m^{-3} and tension units from N m^{-1} to mN m^{-1}. For the lens magnification ($\times$ 2.5) and instrument (RED, site 04) used, x is 1.79×10^{-3} and thus $A = 1.572 \times 10^{-8}$.

$$A = 0.25 \times 0.10472^2 \times x^3 \times 10^6 \tag{1.51}$$

The substitution of the various quantities into Equation (1.50) gives ~1.75 mN m^{-1} as the interfacial tension value.

Microtensiometry is used for investigating interfaces on very small particles and in finely dispersed systems, commonly encountered in criminology, biology and pharmaceutical processing and other related fields where the material quantities of interest are too small to apply conventional tensiometric techniques. The micropipette technique is a typical example of microtensiometry. This technique is very similar to the bubble pressure method because it also depends on the pressure difference across curved interfaces. It was developed to measure the surface tension of microdrops directly and was first used in the study of vesicles (Evans and Needham 1980). The technique involves capturing a liquid drop at the tip of a glass micropipette of radius r_p and then sucking it into the pipette as illustrated in Figure 1.9b (i). The surface tension is then calculated from the minimum pressure necessary for the drop to extend a hemispherical protrusion into the pipette by using the form of the Laplace equation given in (1.52). The ΔP stands for the pressure difference across the liquid and air phases while r_o represents the radius of the exterior segment of the drop. The r_o is typically larger than r_p.

$$\Delta P = 2\gamma_{la} \left(\frac{1}{r_p} + \frac{1}{r_o} \right) \tag{1.52}$$

A relatively large pressure difference is usually required to suck the drop into the pipette when it does not wet or adhere to its glass surfaces, and this is a major drawback of the technique. To circumvent this, a two-pipette technique, where a separation force is applied between the pipettes to deform the drop [Figure 1.9b (ii)], is used. The separation force is measured and the surface tension is computed from the force-drop deformation relation.

In atomic force microscopy, nanometre-sized particles are imaged and the interaction force between them is measured. This technique makes it possible to study roughness, heterogeneity, and interaction forces at the sub-microscopic level and down to molecular level in some cases. Even though it is yet to be established as a technique for measuring interfacial tension directly, its potential use for this at the microscopic and sub-microscopic levels is great. Presently, it is considered as a means of measuring the wetting properties of colloidal particles. The atomic force microscope measures the interaction forces between a cantilever tip and a specimen based on the deflection of the cantilever, which has a known force constant. The cantilever deflection is detected by the reflection of a laser beam. The movement of the specimen under the cantilever tip, both in the horizontal plane for scanning and vertically for force measurement, is controlled by a piezoelectric specimen stage as illustrated in Figure 1.9c (left). Reverse systems with the piezoelectric stage attached to the cantilever holder are also available. With the atomic force microscope (AFM), interaction forces as small as 10^{-12} N can be measured. Measurement of the interfacial tension between the probe tip and a microscopic liquid drop with the AFM has also been described. The capillary forces exerted on the probe tip by the liquid are measured as the tip is inserted into the liquid drop, withdrawn and detached from it as described in Figure 1.9c (right). The calculation of interfacial tension from these force-distance curves depends on the shape of the probe tip. The mathematical equation used for the calculation is similar to that used in a classical macro-detachment technique. One major challenge of using the AFM for interfacial tension measurement is the fabrication of appropriate probe tips. Cylindrical tubes and spheres are often used. Carbon nanotubes can also be used (Dai *et al.* 1996). The accuracy and suitability of the various methods of interfacial tension measurement are summarized in Table 1.3. It is very crucial to note that surface and interfacial tensions are influenced by temperature, as illustrated in Figure 1.10 for water. This stresses the importance of reporting the temperature at which these measurements have been made.

TABLE 1.3

Accuracy and Suitability of Various Methods of Measuring Interfacial Tension

Method	Accuracy/ mN m⁻¹	Surfactant Solutions	Two-Liquid Systems	Viscous Liquids	Melted Metals
			Suitability		
Wilhelmy plate	~0.1	Limited	Good	Very good	Not recommended
du Noüy ring	~0.1	Limited	Reduced accuracy	Not recommended	Not recommended
Maximum bubble pressure	0.1–0.3	Very good	Very good	Not recommended	Yes
Capillary rise	<< 0.1	Very good	Very good, experimentally difficult	Not recommended	Not recommended
Drop volume	0.1–0.2	Limited	Good	Not recommended	Yes
Pendant drop	~0.1	Very good	Very good	Not recommended	Yes
Sessile drop	~0.1	Good	Very good	Very good	Yes

Source: Drelich, J. *et al.*, Measurement of interfacial tension in fluid-fluid systems, *Encyclopaedia of Surface and Colloid Science*, Marcel Dekker, New York, pp. 3152–3166, 2002.

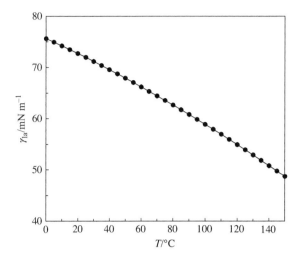

FIGURE 1.10 Temperature T *versus* surface tension γ_{la} for water showing a decrease in the latter as the former increases. (Plotted data taken from Vargaftik, N.B. *et al.*, *J. Phys. Chem. Ref. Data*, 12, 817–820, 1983.)

1.4   CONCLUSION

The concepts of interface and surface tension, which are very important in colloid and surface chemistry, have been clearly explained. Thereafter, the consequences of liquid surface tension, capillarity and wicking, the existence of bubbles and drops, and the walking of insects on liquid surfaces are described. This is followed by a detailed description of the various methods of surface tension measurement. Finally, solved examples along with end-of-chapter questions are given to help the reader appreciate the use of some of the mathematical equations derived or given in the chapter. This chapter is fundamental to the rest, and the reader is encouraged to understand it fully before proceeding to the others.

EXERCISES

DISCUSSION QUESTIONS

Question 1
1. What is an interface?
2. What is an interphase?
3. What is surface tension?

Question 2
1. Differentiate between *capillarity* and *wicking.*
2. Differentiate between *bubbles* and *drops.*
3. Discussion the origin of surface tension in liquids.
4. Discuss the various ways through which surface tension can be measured.

NUMERICAL QUESTIONS

Question 1
1. Derive expressions for the Laplace pressure of the following systems:
   a. an air bubble in water, characterized by one surface, and
   b. a soap bubble, characterized by two surfaces.
2. The surface area of a soap bubble (surface tension = 25 mN m^{-1}) is ~3.14 µm^2. Calculate the pressure difference across the bubble surfaces.

Question 2
1. Derive the Kelvin equation and state its consequences.
2. The vapor pressure, surface tension, and density of methanol at 25°C are 16.9 kPa, 22.07 mN m^{-1}, and 0.7914 g cm^{-3}, respectively. Calculate the volume of a methanol drop whose vapor pressure at 25°C is 30 kPa.

Question 3
What is the corrected value of the surface tension of perfluorohexane (12.8 mN m^{-1}) as measured at 20°C using the du Nöuy ring (radius of ring 9.53 mm and radius of wire 0.20 mm) if the density of air and perfluorohexane at this temperature are 0.0098 and 1.71 g cm^{-3}, respectively?

Question 4
Ultrapure water at 25°C and of density 997 kg m^{-3} was seen to rise 14.7 cm in a capillary tube of inner radius 0.1 mm. Calculate the surface tension of the water at this temperature.

FURTHER READING

Adamson, A.W. and A.P. Gast. *Physical Chemistry of Surfaces.* 6th ed. New York: John Wiley & Sons, 1997.

Barnes, G.T. and I.R. Gentle. *Interfacial Science: An Introduction.* 2nd ed. New York: Oxford University Press, 2011.

Berg, J.C. *An Introduction to Interfaces and Colloids: The Bridge to Nanoscience.* Hackensack, NJ: World Scientific, 2009.

Everett, D.H. *Basic Principles of Colloid Science.* London, UK: Royal Society of Chemistry, 1992.

Hiemenz, P.C. and R. Rajagopalan. *Principles of Colloid and Surface Chemistry.* 3rd ed. New York: Marcel Dekker, 1997.

Pashley, R.M. and M.E. Karaman. *Applied Colloid and Surface Chemistry.* Chichester, UK: John Wiley & Sons, 2004.

Shaw, D.J. *Introduction to Colloid and Surface Chemistry.* 4th ed. Oxford, UK: Butterworth-Heinemann, 1992.

Shchukin, E.D., A.V. Pertsov, E.A. Amelina and A.S. Zelenev. *Colloid and Surface Chemistry.* Amsterdam, the Netherlands: Elsevier, 2001.

REFERENCES

Andreas, J.M., E.A. Houser and W.B. Tucker. "Boundary Tension by Pendant Drops." *J. Phys. Chem.* 42 (1937): 1001–19.

Bashforth, F. and J.C. Adams. *An Attempt to Test the Theories of Capillary Action.* Cambridge, UK: Cambridge University Press, 1883.

Bikerman, J.J. *Physical Surfaces.* New York: Academic Press, 1970.

Binks, B.P., P.D.I. Fletcher, B.L. Holt, P. Beaussoubre and K. Wong. "Phase Inversion of Particle-Stabilized Perfume Oil-Water Emulsions: Experiment and Theory." *Phys. Chem. Chem. Phys.* 12 (2010): 11954–66.

Coucoulas, L.M., R.A. Dawe and E.G. Mahers. "The Refraction Correction for the Spinning Drop Interfacial Tensiometer." *J. Colloid Interface Sci.* 93 (1983): 281–84.

Dai, H., J.H. Hafner, A.G. Rinzler, D.T. Colbert and R.E. Smaley. "Nanotubes as Nanoprobes in Scanning Probe Microscopy." *Nature* 384 (1996): 147–51.

Drelich, J., F. Ch and C.L. White. Measurement of interfacial tension in fluid-fluid systems. *Encyclopaedia of Surface and Colloid Science*, pp. 3152–66. Marcel Dekker: New York, 2002.

Ehlers, W. and M. Goss. *Water Dynamics in Plant Production.* Cambridge, UK: CABI, 2003.

Evans, E. and D. Needham. *Mechanics and Thermodynamics of Biomembranes.* Boca Raton, FL: CRC Press, 1980.

Fordham, S. "On the Calculation of Surface Tension from Measurements of Pendant Drops." *Proc. Roy. Soc. London. Series A Math. Phys. Sci.* 194 (1948): 1–16.

Harkins, W.D. and F.E. Brown. "The Determination of Surface Tension (Free Surface Energy), and the Weight of Falling Drops: The Surface Tension of Water and Benzene by the Capillary Height Method." *J. Am. Chem. Soc.* 41 (1919): 499–525.

Hu, D.L., B. Chan and J.W.M. Bush. "The hydrodynamics of water strider locomotion." *Nature* 424 (2003): 663–66.

Jasper, J.J. "The Surface Tension of Pure Liquid Compounds." *J. Phys. Chem. Ref. Data* 1 (4) (1972): 841–1009.

McCaughan, J.B.T. "Capillarity: A Lesson in the Epistemology of Physics." *Phys. Educ.* 22 (1987): 100–106.

Niederhauser, D.O. and F.E. Bartell. "Report of Progress-Fundamental Research Occurrence and Recovery of Petroleum." Baltimore, MD: American Petroleum Institute, 1948–1949.

Roe, R.-J., V.L. Bacchetta and P.M.G. Wong. "Refinement of Pendant Drop Method for the Measurement of Surface Tension of Viscous Liquids." *J. Phys. chem.* 71 (13) (1967): 4190–93.

Rusanov, A.I. and V.A. Prokhorov. *Interfacial Tensiometry.* Amsterdam, the Netherlands: Elsevier, 1996.

Sottmann, T. and R. Strey. "Ultralow Interfacial Tensions in Water-*n*-Alkane-Surfactant Systems." *J. Chem. Phys.* 106 (20) (1997): 8606–15.

Stauffer, C.E. "The Measurement of Surface Tension by the Pendant Drop Technique." *J. Phys. Chem.* 69 (6) (1965): 1933–38.

Sugden, S. "XCVII.-The Determination of Surface Tension from the Maximum Pressure in Bubbles." *J. Chem. Soc.* 121 (1922): 858–66.

Tabor, D. *Gases, Liquids and Solids and Other States of Matter.* 3rd ed. Cambridge, UK: Cambridge University Press, 1991.

Taylor, H.J. and J. Alexander. "The Measurement of Surface Tension by means of Sessile Drops." *Proc. Industrial Acad. Sci. A* 19 (3) (1944): 149.

Vargaftik, N.B., B.N. Volkov and L.D. Voljak. "International Tables of the Surface Tension of Water." *J. Phys. Chem. Ref. Data* 12 (1983): 817–20.

Vold, R.D. and M.J. Vold. *Colloid and Interface Chemistry.* London: Addison-Wesley, 1983.

Vonnegut, B. "Rotating Bubble Method for the Determination of Surface and Interfacial Tensions." *Rev. Sci. Instr.* 13 (1) (1942): 6–9.

Young, T. *Miscellaneous Works.* Vol. I. London, UK: J. Murray, 1855.

2 Nature of Solid Surfaces

2.1 PARTICLE-LIQUID-AIR THREE-PHASE CONTACT ANGLE AND WETTING

Three interfacial tension forces, namely the solid-liquid (γ_{sl}), liquid-air (γ_{la}), and solid-air (γ_{sa}), act on a liquid drop once placed on the surface of a solid substrate, giving rise to a particle-liquid-air three-phase contact angle θ' (measured in the liquid phase) as illustrated in Figure 2.1a.

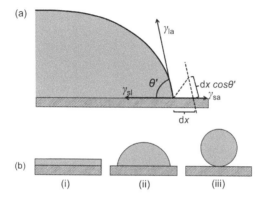

FIGURE 2.1 (a) Illustration of solid-liquid (γ_{sl}), liquid-air (γ_{la}), and solid-air (γ_{sa}) interfacial tension forces acting on a liquid drop placed on the surface of a relatively smooth solid substrate, giving rise to a three-phase contact angle θ'. The drop moves through a distance dx before attaining equilibrium and the θ' becomes θ. (b) Illustration of the three wetting states: (i) complete wetting, (ii) partial wetting, and (iii) dewetting or drying states.

The angle θ' lies between the solid surface and the tangent to the drop surface at the line of contact with the solid surface and moves in response to the interfacial tensions until an equilibrium position is attained. The work done dW by moving a contact line through a distance dx is a product of the net horizontal force F_{net}, Equation (2.1), and the distance dx moved. When the drop is in equilibrium, d$W = 0$ and Equation (2.2) yields Young's law, Equation (2.3) provided the solid surface is relatively smooth. The θ is the equilibrium contact angle.

$$F_{net} = \gamma_{sa} - \gamma_{sl} - \gamma_{la} \cos \theta' \tag{2.1}$$

$$dW = F_{net} \times dx = \underbrace{(\gamma_{sa} - \gamma_{sl})dx}_{\text{from contact line motion}} - \underbrace{\gamma_{la} \cos \theta.dx}_{\text{from creating a new interface}} = 0 \tag{2.2}$$

$$\cos \theta = \frac{\gamma_{sa} - \gamma_{sl}}{\gamma_{la}} \tag{2.3}$$

For a given surface, the equilibrium contact angle can be obtained from the three interfacial tensions. For example, if $\gamma_{sa} = \gamma_{sl} + \gamma_{la}$, $\theta = 0°$ and the system attains equilibrium when a macroscopic uniform liquid layer covers the solid surface leading to a complete wetting state. For the case where $\gamma_{sa} < \gamma_{sl} + \gamma_{la}$, $\theta < 180°$ and partial wetting is witnessed. However, when $\gamma_{sa} \ll \gamma_{sl} + \gamma_{la}$, $\theta \approx 180°$ and the dewetting or the completely dry state is observed. These three scenarios are illustrated in Figure 2.1b. In practice, the drying state is very rare (except for mercury on glass) because van der Waals forces tend to thin vapor layers. "Dry" as used here must not be confused with evaporation where a liquid becomes gaseous without bubble formation (*cf.* boiling). However, it is used to stress the absence of a liquid layer on the solid surface. From a thermodynamic standpoint, the wetting and the dewetting states are very similar, the only difference being that the liquid and vapor phases are interchanged.

An equilibrium spreading coefficient S_e,

$$S_e \equiv \gamma_{sa} - (\gamma_{sl} + \gamma_{la}) = \gamma_{la}(\cos\theta - 1) \tag{2.4}$$

which represents the difference between the solid-air interfacial energy γ_{sa} (also surface energy) and its value in the case of complete wetting, is defined to help classify the wetting states.

Generally, $S_e \leq 0$, and for the case of complete wetting, $S_e = 0$. A drop is normally far from the equilibrium upon arriving the surface and an initial spreading coefficient S_i,

$$S_i = \gamma_{so} - (\gamma_{sl} + \gamma_{la}) \tag{2.5}$$

must be defined. The γ_{so} is the surface energy of the dry solid surface. When $S_i < 0$, the drop acquires a finite contact angle θ_i and prefers to remain in the partial wetting state if it is volatile. For non-volatile liquids and $S_i > 0$, the drop tends to flatten as it attempts to spread out, while preserving its volume (within the experimental time scale). Such drops prefer to wet the surface completely.

The equilibrium γ_{sa} is less than γ_{so} due to the adsorption of molecules on the solid from the vapor phase. This is clear from the analysis of the Gibbs adsorption equation which shows that for high adsorption of gas molecules, the equilibrium γ_{sa} is lowered by $k_B T$ per adsorbed molecule per unit area. The two-dimensional ideal gas law, Equation (2.6), is obtained for the case of low adsorption.

$$\Pi \equiv \gamma_{so} - \gamma_{sa} = k_B T \Gamma \tag{2.6}$$

The Π is a positive "surface force", a force per unit length and Γ is the adsorption. This means that S_i is always more than S_e. The S_i is either positive or negative. Whenever $S_i < 0$, S_e is also negative and the liquid drop does not spread. It is not clear whether $S_e \leq 0$ when $S_i > 0$, but generally, when S_i is large and positive, the equilibrium state is characterized by $S_e = 0$.

There are few exceptions, however. For example, a benzene liquid drop placed on an aqueous substrate spreads rapidly for few seconds and then retracts into a drop again, with a nonzero equilibrium contact angle. This indicates that S_i is positive and S_e is negative. Because benzene is slightly soluble in water, γ_{sl} is low favoring complete wetting but this effect is overpowered by the formation of a microscopic film of the drop phase of the substrate in equilibrium. The benzene layer decreases γ_{sa} significantly and S_e becomes negative as a result. This indicates that S_e cannot be determined in general from the surface tensions of the pure substances involved.

For solid surfaces, wettability is evaluated by drop deposition and measurement of the equilibrium contact angle (Kabza *et al.* 2000). According to Zisman, the difference $\gamma_{sa} - \gamma_{sl}$ is a property of a solid surface, *i.e.* a constant independent of the liquid used, and can be readily obtained by measuring the equilibrium contact angle of a liquid drop of known surface tension on the solid surface, in line with Equation (2.3). The constant is the intercept of the straight line with the horizontal $\cos\theta = 1$. This is the critical surface tension (Fox and Zisman 1950), the tension below which wetting occurs.

It is very difficult to measure the initial spreading coefficients for liquids on solids. One way around this problem is to measure the adsorption isotherm of the liquid on a solid and use the Gibbs adsorption equation to calculate the surface tension change associated with the adsorption (Ragil *et al.* 1996).

2.1.1   MEASUREMENT OF PARTICLE-LIQUID-AIR CONTACT ANGLES FOR SURFACES AND POWDERED PARTICLES

It is easier to measure the particle-liquid-air contact angle θ of macro-surfaces compare to powdered particles (micro- and nanoparticles) because factors like line tension, surface inhomogeneity, and particle size become important at the nanoscale level. In this subsection, common methods for the measurement of θ for both macro-surfaces and powdered particles will be considered. Many of these methods involve growing (or expanding) a liquid drop on a solid substrate followed by measuring the tangent angle at the three-phase contact line when equilibrium is attained. Such angles are known as advancing contact angles. This can be compared to receding contact angles, measured from a contracted sessile liquid drop by reducing its volume.

2.1.1.1   Measurement by Telescope-Goniometer

This is the most widely used method of contact angle measurement. It is a direct method which involves measuring the tangent (particle-liquid-air) angle at the three-phase contact point on a sessile drop profile resting on a planar surface. The telescope-goniometer was invented by Bigelow *et al.* (1964). However, the first commercial one was designed by W.A. Zisman and was manufactured by Ramé-Hart Instrument Company. The instrument, Figure 2.2a, is made up of a horizontal stage for mounting solid substrates or liquid samples, a micrometer pipette for forming a liquid drop, an illuminating source and a telescope equipped with a protractor eyepiece. Measurement is done by simply aligning the tangent of the sessile drop profile at the contact point with the surface and reading the protractor through the eyepiece. Following modifications to improve accuracy and precision, many of the modern ones are integrated with a camera to take photographs of the drop profile so

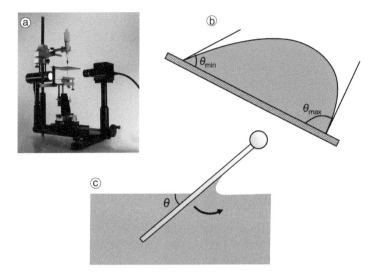

FIGURE 2.2 (a) Photograph of a Ramé-hart contact angle goniometer, (b) illustration of the "titled plate" method of contact angle measurement showing the θ_{max} and θ_{min}, which correspond to the advancing and receding angles, respectively, and (c) illustration of the tilting plate method.

as to measure the angles at leisure. The use of a relatively high magnification enables a detailed examination of the intersection profile (Smithwich 1988). A motor-driven syringe is sometimes used to control the rate of liquid addition and removal so as to study the advancing, receding, or dynamic contact angles (Kwok *et al*. 1996). This direct optical method is advantageous for three reasons. It is simple and requires (a) only a small amount of liquid (few microliters) and (b) small substrate (few square millimetres). Unfortunately, the impact of impurities is quite high due to the small size of the drop and substrate surface. In addition, the degree of accuracy and reproducibility of the measurement depends on the consistency of the operator in assigning the tangent line. Also, the imaging device only focuses on the largest meridian section of the sessile drop and, as a result, the profile image reflects only the contact angle at the point the meridian plane intersects the three-phase contact line.

Surface inhomogeneity or roughness also varies the contact point along the three-phase contact line. Another limitation of this method is that small contact angles (<20°) cannot be measured accurately because of the difficulty in assigning a tangent line when the drop profile is almost flat. Lastly, contact angle values depend on drop size and this causes a systematic problem. Despite these drawbacks, this method is considered as the most convenient way of contact angle measurement if high accuracy is not required. The accuracy of this method is about ±2° (Neumann and Good 1979).

To obtain good angles, it is recommended that the telescope be tilted down slightly (1° to 2°) of the horizon so as to bring the drop profile reflected by the substrate surface into focus, eliminating the formation of a fuzzy drop-substrate contact line in the profile. Also, the background light required for observation should be one that would not heat the drop or the substrate. To obtain an advancing contact angle, the sessile liquid drop is slowly grown to a diameter of about 5 mm using a micrometre syringe with a narrow gauge stainless steel or Teflon needle. The needle diameter is required to be as small as possible so that it does not distort the drop profile shape. To avoid unwanted vibrations, the needle remains in the drop during the measurement. Because the drop may be unsymmetrical, it is advisable to take an average of the angles on both sides of the drop. For a relatively large surface, an average of the angles at multiple points should be taken to obtain a true representative of the entire surface. Measurements should be made inside an enclosed chamber to exclude air-borne contamination and establish an equilibrium vapor pressure of the test liquid, especially if it is volatile (Yuan and Lee 2013). The receding angles are obtained, similarly, after contracting the drop by withdrawing some liquid into the syringe.

The difference between the advancing and receding angles is known as contact angle hysteresis. The measurement of contact angle hysteresis has been recommended as a means of assessing the quality of a solid substrate. Macdougall and Ockrent (1942) modified the sessile drop method by tilting the solid substrate until the drop just begins to move to obtain both the advancing and the receding contact angles. This method is known as the tilted plate or inclined plate method. This method must not be confused with the tilting plate method discussed in Section 2.1.1.2. The contact angles obtained at the lowest point θ_{max} and the highest point θ_{min}, Figure 2.2b, are considered as the advancing and receding contact angles, respectively. This method has been used by Extrand and Kumagai (1996) and Extrand and Kumagai (1997) in their study of contact angle hysteresis of liquids on different polymer surfaces, including silicon wafers and elastomeric surfaces. Nonetheless, the linkage between the advancing/receding angles with the maximum/minimum contact angles must be used with caution because they are quite different in some cases (Krasovitski and Marmur 2005, Pierce *et al.* 2008).

2.1.1.2   Tilting Plate Method

This method was developed by Adam and Jessop in 1925. It is simple, and the contact angle values depend less on the operator's subjectivity. Measurement consists of gripping one end of a plate above a liquid and rotating it toward the liquid surface until the end of the plate is immersed in the liquid [Figure 2.2c]. A meniscus is formed on both sides of the plate. The plate is then tilted slowly until the meniscus becomes horizontal on one side of the plate. The angle between the plate and the horizontal becomes the desired contact angle. Adam and Jessop reported an error of $\pm 5°$ but attributed it to liquid contamination. The requirement of considerable skills and the disturbance of the liquid by the rotating plate are the major difficulties associated with the method. The accuracy of this method was improved by Fowkes and Harkins (1940) through the use of barriers for surface cleaning and a film balance to detect the presence of impurities on the liquid surface and a microscope. The tilting plate method has been used to measured small contact angles ($<10°$). The precision and reproducibility of this method has been improved by Bezuglyi *et al.* (2001) through the application of a high sensitive thermocapillary response to the static curvature of the liquid meniscus.

2.1.1.3   Captive Bubble Method

The captive bubble method of contact angle measurement was introduced by Taggart *et al.* (1930). It can be thought of as the opposite of the sessile drop technique. Rather than forming a sessile drop on a solid substrate, it is immersed in the test liquid, and an air bubble is formed beneath it. Usually, a small amount of air (~0.05 mL) is injected into the liquid of interest to form the air bubble. Similar to the sessile drop method, the needle is held in the bubble throughout the measurement to avoid disturbing the balance of the advancing angle and to prevent the bubble from drifting over the solid surface if the substrate is not perfectly horizontal.

One advantage of this method is that the substrate is in contact with a saturated atmosphere. The risk of solid-air interface contamination from airborne oil drops is also minimal with the method. In addition, it is much easier to monitor the temperature of the liquid in the captive bubble method than with sessile drops, making it possible to investigate the temperature dependence of contact angles. There is good agreement between the captive bubble and sessile drop methods for smooth polymeric surfaces (Zhang *et al.* 1989). One drawback of the method is that it requires far more volume of liquid than the sessile drop method. Also, the method is limited in the cases where the substrate swells upon immersion in the liquid or a surface film is dissolved by the liquid.

2.1.1.4 The Wilhelmy Balance Method

This is an indirect method of contact angle measurement. The contact angle value is obtained from the change in weight, detected by a balance, when a thin, smooth, and vertical plate (*i.e.* solid substrate) is brought in contact with the test liquid as illustrated in Figure 2.3 (upper image). The change in weight is a consequence of a combination of buoyancy and wetting forces acting on the plate, in the presence of

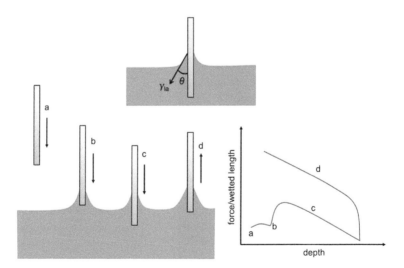

FIGURE 2.3 The upper image shows the illustration of the Wilhelmy balance method. The lower image shows the illustration of the submersion cycle for the Wilhelmy balance measurement showing: (a) the approach of the liquid by the solid substrate where the force/wetted length is zero, (b) first contact of the solid substrate with the liquid leading to θ < 90°, the liquid rises up, causing a positive wetting force, (c) further immersion of the substrate which increases the buoyancy and decreases the force detected on the balance, and (d) the pulling out of the substrate from the liquid after the desired depth is reached. The force in (c) is measured for the advancing angle while that in (d) is measured for the receding angle.

a constant force of gravity, once in the liquid. The wetting force f_w, which acts vertically downward on the plate, is

$$f_w = \gamma_{la} X \cos\theta \qquad (2.7)$$

in which γ_{la} is the liquid-air interfacial (or surface) tension of the test liquid, X is the perimeter of contact line (which is equal to the perimeter of the substrate's cross-section), and θ is the contact angle. The buoyancy force f_b which acts vertically upward is

$$f_b = V\Delta\rho g \qquad (2.8)$$

The V is the volume of liquid displaced, $\Delta\rho$ is the difference in density between the liquid and air and g is the acceleration due to gravity. The overall force change ΔF, detected by the balance, is

$$\Delta F = \gamma_{la} X \cos\theta - V\Delta\rho g \qquad (2.9)$$

The contact angle value is readily obtained from Equation (2.9) once the liquid surface tension, substrate perimeter, volume of liquid displaced, and the difference in density between the liquid and air are known. In few cases, when θ is zero and X is known, ΔF is directly related to the liquid surface tension. In a technique developed by Princen (1969), measurement of a zero-contact angle value is possible, therefore permitting the measurement of liquid surface tension by the Wilhelmy balance method.

Example 2.1: Calculating Contact Angle from the Wilhelmy Balance Experiment

Given that the density and surface tension of water at 25°C are approximately 997 kg m^{-3} and 72 mN m^{-1}, calculate the contact angle in a Wilhelmy balance experiment where the plate perimeter is 36.2 mm and the overall force change is 2.63 mN while the volume of water displaced is 1×10^{-4} cm^3 and the density of air at the prevailing temperature is 1.184 kg m^{-3}.

Method
Substitute the given quantities into Equation (2.9) and solve for the contact angle θ.

Answer
From Equation (2.9),

$$\cos\theta = \frac{\Delta F + V\Delta\rho g}{\gamma_{la} X}$$

$$= \frac{(2.63\times10^{-3})\,N + [(1\times10^{-10})\,m^3 \times (997 - 1.184)\,kg\,m^{-3} \times 9.8\,m\,s^{-2}]}{72\times10^{-3}\,N\,m^{-1} \times 36.2\times10^{-3}\,m}$$

$$\approx 1.0$$

$$\Rightarrow \theta = \cos^{-1}(1.0) = 0°$$

The advancing and receding contact angles are obtained by pushing into and pulling out the solid substrate from the liquid, respectively. The overall process is illustrated in Figure 2.3 (lower image): (a) the force/length is zero as the solid substrate approaches the liquid-air interface, (b) the substrate makes contact with the liquid-air interface, a contact angle $\theta < 90°$ forms, the liquid rises up and a positive wetting force develops, (c) the force detected decreases as buoyancy increases upon further immersion of the substrate, this force is measured for the advancing angle, and (d) the substrate reaching the desired depth is pulled out of the liquid, the force is measured for the receding angle.

The Wilhemy balance method has numerous advantages compared to conventional optical methods (Yuan and Lee 2013). First, the hard work of contact-angle measurement is reduced to the measurement of weight and length, which can be done with high accuracy and without subjectivity. Second, the value of force obtained at any given depth of immersion is an average value, a feature which automatically gives a more accurate contact angle value that represent the property of the substrate surface. Unfortunately, it does not help determine surface heterogeneity. Third, the graph obtained during measurement can be used in dynamic contact angles and contact angle hysteresis studies at different wetting speeds. In addition, the smoothness of the curve is related to the heterogeneity of the solid substrate. Lastly, absorption or surface reorientation can be studied simply by repeating the submersion circle.

This method is also associated with some drawbacks. The solid substrate is required to have a uniform cross section in the submersion direction. Although rods, plates, and fibres of known perimeters are ideal samples, it is sometimes difficult to accurately measure the perimeter and the wetted length. Apart from the regular geometry requirement, the substrate is required to have the same composition and topography at all sides. This is difficult to achieve especially in the case of film investigation or anisotropic systems. Finally, the method requires large quantity of liquid and the chance that the substrate will swell and/or absorb vapor is very high.

2.1.1.5   Capillary Rise at Vertical Plate

Due to capillary effect, a liquid rises once in contact with a vertical and wide (2 cm) plate. The rise height h is related to the contact angle θ by

$$\sin\theta = 1 - \frac{\Delta\rho g h^2}{2\gamma_{la}} \tag{2.10}$$

with $\Delta\rho$ being the difference in density between liquid and air, g being the acceleration due to gravity and γ_{la} being the liquid-air interfacial tension. The Wilhelmy balance method can also be modified to measure the capillary rise and hence θ. Measurement of dynamic contact angle is achieved by moving the plate up or down. This method has been widely used and has been shown to be especially suitable for studying the dependence of contact angles on temperature. It has been automated by Budziak and Neumann (1990) and Kwok *et al.* (1995). An accuracy of $\pm 0.1°$ can be obtained with this method if the surface is well prepared and forms a straight meniscus line. In addition to contact angle measurement, it is also possible to obtain the surface tension of the test liquid by combining Equations (2.9) and (2.10) *via* $\sin^2\theta + \cos^2\theta = 1$ (Jordan and Lane 1964). The method has similar advantages and disadvantages to the Wilhelmy balance method.

Example 2.2: Calculating Contact Angle from Capillary Rise at a Vertical Plate

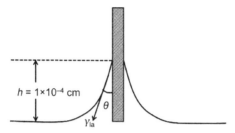

Suppose that the plate in Example 2.1 is replaced with a vertical glass one and a rise height $h = 1 \times 10^{-4}$ cm was observed, calculate the contact angle θ.

Method

Substitute the necessary quantities into Equation (2.10) and solve for the contact angle θ.

Answer

$$\sin\theta = 1 - \frac{\Delta\rho g h^2}{2\gamma_{la}} = 1 - \frac{(997 - 1.184)\ \text{kg m}^{-3} \times 9.8\ \text{m s}^{-2} \times (1 \times 10^{-6})^2\ \text{m}^2}{2 \times 72 \times 10^{-3}\ \text{N m}^{-1}} \approx 1$$

$$\Rightarrow \theta = \sin^{-1}(1) = 90°$$

2.1.1.6 The Capillary Tube Method

The meniscus of a sufficiently narrow tube inserted vertically in a bulk liquid (Figure 1.6) is considered to be approximately spherical. The relationship between the capillary rise h, the capillary radius r, the density difference between the liquid and vapor phases $\Delta\rho$, surface tension γ_{la}, and the contact angle θ is

$$h = \frac{2\gamma_{la}\cos\theta}{\Delta\rho g r} \tag{2.11}$$

in which g is the acceleration due to gravity. This relationship is known as the Jurin rule, named after James Jurin who studied the effect in 1718. The contact angle is calculated from experimentally measured values of h and r. It is also calculated from the length of the capillary occupied by a known mass of mercury when r is too small.

2.1.1.7 Method of Capillary Penetration for Powders and Granules

The degree of wetting of powders and granules by liquids also depends on the contact angle the liquids make with the individual powdered particles or a single granule. However, it is very difficult to measure the contact angles an individual powdered particle or a single granule makes with liquid phases. Rather, the powders or granules are compressed into flat pellets and the contact angles the liquid drops make with them are measured (Clint and Wicks 2001, Binks *et al.* 2014, Tyowua *et al.* 2017). Unfortunately, due to the inherent porous nature of the particle pellets, liquid penetration often occurs, especially if the contact angle is <90°. This usually affects contact angle data and its reproducibility. Contact angle values of liquid drops on particle pellets are also influenced by surface roughness and particle swelling. In addition, the topmost particles often undergo plastic deformation during compression. This also influences the contact angle values compared to the uncompressed powder. The capillary penetration method, developed by Washburn (1921), is used to surmount these problems. The method involves measuring the depth of the liquid front intrusion *versus* time. The relationship between the depth of liquid intrusion l, the surface tension γ_{la}, viscosity of the liquid η, liquid penetration time t, the pore radius, and the contact angle θ, according to Washburn, is

$$l^2 = \frac{rt\gamma_{la}\cos\theta}{2\eta} \tag{2.12}$$

The Washburn method has undergone several modifications. Bartell and Osterhof (1927) proposed static measurements, which has been further modified by White (1982). The wetting liquid penetrates vertically upward through the compressed powder pellet until it reaches a height at which the capillary pressure is balanced by the weight of the liquid in the column. The Laplace pressure ΔP, which drives the liquid into the pores of radius σ, is measured and used for contact angle calculation using Equation (2.13).

$$\Delta P = \frac{2\gamma_{la}\cos\theta}{\sigma} \tag{2.13}$$

Unfortunately, the pore radius is not uniform throughout the powder bed, and an effective pore radius σ_{eff} must be defined. The expression according to White (1982) is

$$\sigma_{eff} = \frac{2(1-\phi)}{\phi\rho\,A} \tag{2.14}$$

where ϕ stands for the volume fraction of solid in the packed pellet, ρ stands for the density of the solid material, and A is the specific surface area per gram of solid.

The substitution of Equation (2.14) into (2.13) gives the Laplace-White equation (2.15), as applied to a porous media.

$$\Delta P = \frac{\gamma_{la}\phi\rho\ A\cos\theta}{1-\phi} \tag{2.15}$$

There is a good agreement between this equation and experimental data for clean and chemically treated smooth glass beads. Nonetheless, it is very difficult to determine σ_{eff} directly due to the uncertainty in measuring the specific area the liquid wets the particles. As a result many researchers (Dunstan and White 1986, Diggins *et al.* 1990) have used a second (reference) liquid of low surface tension that completely wets the powdered particles and then compare it with the wetting behavior of the test liquid. The success of this method depends on the right choice of reference liquid. Although the contact angles obtained on powdered particle surfaces are not identical to Young's contact angles, the actual measured angles, together with the surface tension of the contacting liquid, determine the Laplace pressure. That is, it is the actual contact angle that determines the wetting behavior of the system.

2.1.1.8   Drop Shape Analysis

There are numerous methods for measuring liquid surface tension as well as contact angle from the shape of sessile liquid drops, pendant liquid drops, or captive bubbles. Recall that the shape of a liquid drop depends on the combined effects of interfacial tension and gravitational forces. Interfacial tension minimizes the surface area by making the drop spherical while gravity deforms the drop through elongation (pendant) or flattening (sessile). The balance between the interfacial tension and the gravitational forces is sum-up in the Laplace Equation (1.33 or 1.45) of capillarity, which makes determination of interfacial tension and contact angles *via* the analysis of drop shape possible.

Geometrically, the contact angle is calculated from the knowledge of the drop radius R_{drop} and the height h of the drop apex using Equation (2.16).

$$\frac{\theta}{2} = \tan^{-1}\left(\frac{h}{R_{drop}}\right) \tag{2.16}$$

This method gives good contact angle values when the liquid drop is extremely small, which is considered to be spherical. Because the gravitational forces dominate interfacial tension forces in the case of large liquid drops, these drops are non-spherical as a result of deformation and Equation (2.16) is inapplicable.

Example 2.3: Calculating Contact Angle from Analysis of Drop Shape

A quasi-spherical drop of height 3.47 mm and diameter 3.55 mm, *cf.* Figure 3.4a, is resting on a smooth planar substrate in air. Calculate the contact angle the drop makes with the substrate.

Method

Substitute the given quantities into Equation (2.16) and solve for the contact angle θ.

Answer

$$\frac{\theta}{2} = \tan^{-1}\left(\frac{h}{R_{drop}}\right) = \tan^{-1}\left(\frac{3.47 \text{ mm}}{0.5 \times 3.55 \text{ mm}}\right) = 63°$$

$$\Rightarrow \theta = 2 \times 63° = 126°$$

This value can be compared with the experimentally measured value of ~158° for the quasi-spherical drop in Figure 3.4a whose geometry is the same with that described in this example.

2.1.1.9 Axisymmetric Drop Shape Analysis (ADSA)

This method of contact angle measurement was developed by Rotenberg *et al.* (1983) and improved by Spelt *et al.* (1987), Cheng *et al.* (1990), del Rio and Neumann (1997), and Kalantarian *et al.* (2009). The method is believed to be one of the most accurate methods of contact angle measurement. It has a reproducibility of ±0.2° compared to ±2° by direct tangent measurements. This method assumes that the experimental liquid drop is Laplacian and axisymmetric and that gravity is the only external force acting on the drop. By obtaining the best theoretical profile that matches that obtained from an experimental image, the surface tension, contact angle, drop volume, and surface area can be calculated. Surface tension is usually the adjustable parameter and the algorithm chooses the value that gives the best theoretical profile that fits the experimental drop profile.

The axisymmetric drop shape analysis-diameter (ADSA-D) is another type of ADSA method used in the measurement of extremely low contact angles (typically <20°) or contact angle on non-ideal surfaces. Its working principle is different from that described previously. The ADSA-D programme analyzes a top view image of the drop and measures the contact diameter. The contact angle is determined from numerical integration of the Laplace equation of capillarity once the contact diameter, liquid surface tension, and volume of the liquid drop are known.

2.1.2 CONTACT ANGLE HYSTERESIS

A liquid drop placed on a solid surface soon attains a unique equilibrium contact angle known as the Young contact angle θ, provided the surface is smooth. Due to the long-range interaction between molecules of the drop (at the contact line) and molecules of the solid surface, a diversity of contact angles is observed, depending on whether the drop is advancing or retracting (receding). In many cases the angle lies between values of the advancing θ_a and the receding θ_r angles. These angles fall within a range, with the advancing angles approaching a maximum value and the receding angles approaching a minimum value, in many cases. While the experimental advancing contact has high reproducibility, the receding one has less reproducibility due to liquid sorption or solid swelling by the liquid. The difference between the advancing and receding contact angles is known as contact angle hysteresis. The origin of contact angle hysteresis is not known precisely, but it is thought to arise from surface roughness or heterogeneity (chemical, *e.g.* presence of impurities or textural). The rough or impure domains pin the three-point contact line making it energetically unfavorable for it to move (Eral *et al.* 2012) as illustrated in Figure 2.4. This is because an energy barrier U needs to be surmounted

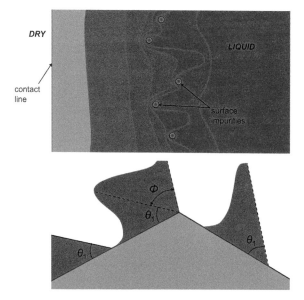

FIGURE 2.4   The upper image shows the pinning of a contact line receding from left to right as a result of surface impurities. The lower image shows the possible contact angles at the corner due to contact line pinning.

by contact line before displacement (Whyman *et al.* 2008). Hysteresis is generally expressed as

$$\theta_a - \theta_r = \left(\frac{8U}{\gamma_{la} R_o} \right)^{0.5} h(\theta) \tag{2.17}$$

where R_o is the initial radius of the spherical drop before deposition on the solid surface and $h(\theta)$ is a function of the equilibrium contact angle and its values are reported by Whyman *et al.* (2008). On rough surfaces, the "phobic" domains tend to pin the motion of the liquid drop as it advances or recedes thereby increasing or decreasing the contact angles, respectively. The opposite occurs for the case of the "philic" domains and the corresponding angles are small. These angles are not necessarily Young contact angle, especially if the surfaces are not smooth. Contact angle hysteresis is not observed on ideal solid surfaces and the experimental contact angle represents the Young contact angle. Smooth but chemically heterogeneous solid surfaces give contact angle that might not represent the Young contact angle, but the advancing contact angle is approximately equal to it.

Although there are no general guidelines describing the degree of surface smoothness, roughness effects on contact angle values cannot be ignored. It is recommended that the solid surface should be made as smooth as possible and should be inert to the test liquid too. Smooth homogeneous surfaces are obtained through heat pressing, solvent casting, self-assembled monolayers, dip coating, vapor deposition, and surface polishing.

Contact angle hysteresis occurs in nature and has many ramifications. The trapping of a liquid column in capillary tube despite the effects of gravity and the pinning of a raindrop on a window pane, car window, or airplane and the sticking of pesticides to leaves are all consequences of contact angle hysteresis (Eral *et al.* 2012). Consider a heavy liquid column trapped in a capillary tube [Figure 2.5a]. The θ_2 can be as large as θ_a while the θ_1 can be as small as θ_r. Generally, $\theta_2 > \theta_1$ and a net capillary force exists, as a result, to support the weight of the liquid column. The force balance requires the maximum contact force to equal the weight of the column as shown in Equation (2.18), which rearranges to (2.19).

$$\underbrace{2\pi R\gamma_{la}(\cos\theta_1 - \cos\theta_2)}_{\text{max contact force}} = \underbrace{\rho g\pi R^2 h}_{\text{weight of column}} \tag{2.18}$$

$$\frac{2\gamma_{la}}{R}(\cos\theta_1 - \cos\theta_2) = \rho gh \tag{2.19}$$

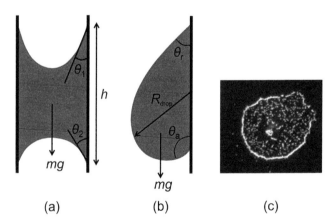

(a)   (b)   (c)

FIGURE 2.5 (a) Illustration of a heavy liquid column trapped in a capillary tube of radius R despite the effects of gravitational force, (b) schematic of a pinned raindrop on a window pane, and (c) microscope image of a coffee stain formed by fluorescently labelled 5 μm particles. (Taken from Eral, H.B. *et al.*, *Colloid Polym. Sc.*, 291, 247–260, 2013.)

Therefore, an equilibrium is possible only if $2\gamma_{la}/R(\cos\theta_r - \cos\theta_a) > \rho gh$. The γ_{la} and ρ are the surface tension and density of the liquid, respectively, while R is the radius of the capillary tube, h is the length of the liquid column and g is the acceleration due to gravity. When $\theta_a = \theta_r$ (i.e. no hysteresis), there is no equilibrium and the liquid column collapses. The differences in contact angles around the perimeter of a drop may result in a net force that resists its motion. Consider a raindrop on a vertical window pane [Figure 2.5b]. The drop's motion in response to the force of gravity $F_g = mg$ is opposed by the resultant resistive contact force F_c. The F_c is proportional to $2\pi R_{drop}(\cos\theta_r - \cos\theta_a)$ with the liquid surface tension γ_{la} as the proportionality constant, i.e. $F_c = 2\pi R_{drop}\gamma_{la}(\cos\theta_r - \cos\theta_a)$ (Hu and Bush 2010). The R_{drop} and m are the drop contact radius and the drop mass, respectively. For $F_c \geq F_g$, the drop remains stuck to the window pane. However, if $F_c < F_g$ it rolls down the window pane.

Contact angle hysteresis also plays a role, which may be desirable or undesirable, in some industrial processes. For example, contact angle hysteresis is a problem in immersion lithography where a liquid drop is required to go to only where it is desired, but it is crucial in dip coating where residual liquid drops are required to spread to other places upon dipping and removing the necessary material from the bulk liquid. Contact angle hysteresis also influences pattern formation, the so-called "coffee stain" phenomenon commonly observed in coffee, seen after evaporation of a liquid drop containing non-volatile components. A liquid drop containing non-volatile components like polymers, microparticles or nanoparticles, biomolecules such as DNA or proteins leaves behind a heterogeneous solid residue [Figure 2.5c] upon evaporation due to contact angle hysteresis and evaporation-driven capillary flow, which tends to pull the contents of the droplet towards the contact line. At the vicinity of the drop periphery, where the drop is thinnest and the contact line is pinned, the rate of liquid evaporation is dramatically higher compared with elsewhere, sweeping the contents to the contact line *via* capillary flow. As the drop evaporates with the contact line pinned, the particles accumulate and jam near it thereby strengthening the pinning. The coffee stain phenomenon is very important in evaporating self-assembly (Harris *et al.* 2007). The coffee stain phenomenon is also important in curtain coating where a large drop of liquid falls onto a moving solid surface (*e.g.* a tape or similar) and coats it as a result. However, coffee stain phenomenon is undesirable in many industrial processes involving non-volatile components dispersed in evaporating solvents, *e.g.* inkjet printing of circuits, OLED displays, and drying of paint, where a homogeneous distribution of the non-volatile component is required. This necessitates the prevention of coffee stain. AC electrowetting has been used to control hysteresis and hence the effects of coffee stain (Li and Mugele 2008 and 't Mannetje *et al.* 2011). This is achieved by rapidly shaking the drop using an electric field that rapidly turns on and off (*i.e.* an AC signal) (Li and Mugele 2008 and 't Mannetje *et al.* 2011).

2.1.3 WETTING STATES

2.1.3.1 Young Wetting State

The wetting of atomically smooth solid surfaces is characterized by an equilibrium or Young contact angle θ, Figure 2.6a, which is related to the solid-liquid, liquid-air, and solid-air interfacial tensions through Equation (2.3). The angle depends on

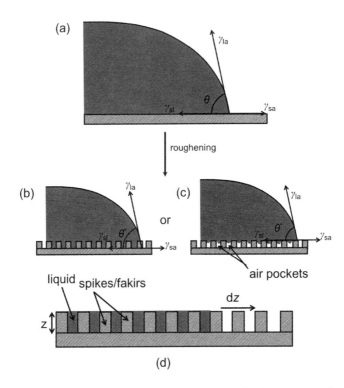

FIGURE 2.6 Schematics of (a) Young, (b) Wenzel, and (c) Cassie-Baxter wetting states and (d) liquid film propagating in geometric textures on a solid surface.

external parameters like temperature. A change in temperature may induce a transition from partial wetting to complete wetting, *i.e.* a wetting transition on a smooth surface would be observed.

2.1.3.2 Wenzel Wetting State

A rough (or textured) surface is a type of 2-D porous material, susceptible to liquid invasion. When a liquid drop is placed on it, a film can propagate inside the roughness so that the drop eventually coexists with it. For a liquid drop placed on a rough solid surface, Figure 2.6b, c, the change in surface energy dE associated with the drop front advancing a distance dz, after rising to a height z as illustrated in Figure 2.6d, is

$$dE = (\gamma_{sl} - \gamma_{sa})(r_s - \phi_s)dz + \gamma_{la}(1 - \phi_s)dz \qquad (2.20)$$

where the r_s and the ϕ_s are roughness parameters defined as

$$r_s = \frac{\text{Total surface area}}{\text{Projected surface area}} > 1 \qquad (2.21)$$

$$\phi_s = \frac{\text{Area of spikes top}}{\text{Total surface area}} < 1 \qquad (2.22)$$

so that

$$\cos\theta_c \equiv \frac{1-\phi_s}{r_s-\phi_s} \qquad (2.23)$$

The first term in Equation (2.20) expresses the fact that the liquid wets the inside of the texture, but not the top (*i.e.* partial wetting), while the second term comes from the fact that the progression of the film leads to the formation of a liquid-air interface. The energy decreases if $\theta < \theta_c$. For flat smooth surfaces ($r_s = 1$), the liquid invades the substrate if the contact angle vanishes ($\theta_c = 0$), which rarely occurs. For porous media ($r_s = \infty$) and the classical condition ($\theta_c = \pi/2$) for impregnation results. For rough surfaces, the condition is intermediate between these two extremes in line with Equation (2.23), which requires θ_c to be intermediate between 0° and 90°. This is logical as hemi-wicking itself is intermediate between complete wetting and usual wicking. By controlling the contact angle through surface chemistry, r_s and ϕ_s with geometry, the wettability of a solid surface can be tuned. The Wenzel state arises when the liquid impregnates a rough chemically homogeneous solid surface as illustrated in Figure 2.6b. The change in wetting energy dE_W associated with the liquid front advancing a distance dz is

$$dE_W = r_s(\gamma_{sl} - \gamma_{sa})dz + \gamma_{la}\cos\theta^*dz \qquad (2.24)$$

The θ^* is the apparent contact angle of the drop on the rough surface. When $r_s = 1$ (smooth surface), the Young equation and its wetting state emerges. However, if $r_s > 1$,

$$\cos\theta^* = r_s\cos\theta \text{ (the Wenzel law)} \qquad (2.25)$$

The phobic and philic characters of solid surfaces, as prescribed by θ, are enhanced by surface roughening, *e.g.* for phobic surfaces $\theta > \pi/2$, $\cos\theta < 0$ and $\theta^* \gg \pi/2$ for large r_s. At large r_s, air is trapped within the rough surface and the Wenzel state breaks down to the Cassie-Baxter state, Figure 2.6c. Drops in the Cassie-Baxter state are said to be in a 'fakir state', while those in the Wenzel state are said to be in 'inverse fakir state'. Intermediate between these states is the 'impaled state', where the rough elements on the solid surfaces spear the liquid drop. These three states are illustrated in Figure 2.7.

2.1.3.3   Cassie-Baxter Wetting State

The wetting of flat, chemically heterogeneous solid surfaces is characterized by an apparent contact angle, predicted by the Cassie-Baxter wetting model (Cassie 1948 and Cassie and Baxter 1944). According to this model, the capillary pressure prevents the invasion of the rough solid surface by the liquid, leaving a trapped vapor layer. In other words, the liquid drop sits on the rough elements (spikes,

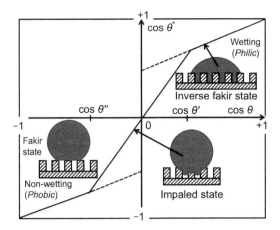

FIGURE 2.7 Schematic of the Wenzel-Cassie-Baxter wetting states diagram showing regions of wetting and non-wetting.

islands or pillars) and the energy change associated with the drop front advancing a distance dz is

$$dE_C = \phi_s(\gamma_{sl} - \gamma_{sa})dz + (1 - \phi_s)\gamma_{la}dz + \gamma_{la}\cos\theta^* dz \qquad (2.26)$$

At equilibrium, $dE_C/dz = 0$ and

$$\cos\theta^* = \phi_s - 1 + \phi_s\cos\theta \qquad (2.27)$$

As $\phi_s \to 0$, $\cos\theta^* \to -1$, *i.e.* $\theta^* \to \pi$. The maintenance of a Cassie-Baxter state is key to water repellency.

2.1.3.4 Transition between the Wenzel and the Cassie-Baxter Wetting States

It is important to note that the Wenzel and the Cassie-Baxter models need to be considered together to completely describe the wettability of a solid surface. As a result, using Equations (2.25) and (2.27), values of $\cos\theta^*$ are plotted against values of $\cos\theta$ in Figure 2.7, where $r_s = 2$ and $\phi_s = 0.4$. Four wetting states are obvious in addition to the metastable states (broken lines) where both the Wenzel and Cassie-Baxter states coexist: (i) the Cassie-Baxter partial wetting state where $\cos\theta' < \cos\theta < 1$, (ii) the Wenzel partial wetting state where $0 < \cos\theta < \cos\theta'$, (iii) the Wenzel non-wetting state where $\cos\theta'' < \cos\theta < 0$, and (iv) the Cassie-Baxter non-wetting state where $-1 < \cos\theta < \cos\theta''$. Whether the Wenzel or Cassie-Baxter wetting state will prevail depends on the magnitudes of dE_W and dE_C. For example, the Wenzel state will give way to the Cassie-Baxter state if $dE_W > dE_C$, but the Wenzel state prevails over the Cassie-Baxter state when $dE_W < dE_C$. The former requires that

$$-r_s\cos\theta + \cos\theta^* > -\phi_s\cos\theta + (1 - \phi_s) + \cos\theta^* \qquad (2.28)$$

$$\cos\theta > \frac{\phi_s - 1}{r_s - \phi_s}, \text{ where } \cos\theta_c = \frac{\phi_s - 1}{r_s - \phi_s}, \Rightarrow \cos\theta > \cos\theta_c \qquad (2.29)$$

This is also the criterion for the occurrence of spontaneous wicking. Similarly, the latter transition requires that $\cos\theta < \cos\theta_c$. Generally, gravity, pressure, drop evaporation, electrowetting, and drop vibration induces wetting transition (Bormashenko 2010). Many wetting transitions are irreversible.

2.2 INTERFACIAL ENERGY OF SOLID SURFACES

The solid-air interfacial energy γ_{sa} of solid surfaces is akin to the liquid-air interfacial tension of liquids, and its unit is the same N m^{-1}. It may be defined as the force per unit length acting perpendicularly to the surface of a solid. It is also defined as the change in the total surface free energy G per unit change in surface area A at constant temperature T, pressure P and moles n, i.e. $\gamma_{sa} = (\partial G/\partial A)_{T,P,n}$ (Chaudhury 1996) and therefore its unit is sometimes expressed in J m^{-2}. The Young Equation (2.3), which describes the force balance between the solid surface energy, liquid surface tension, and solid-liquid interfacial energy when a liquid drop is placed on a solid surface can be used to estimate the value of the solid surface energy provided the other two forces and the equilibrium contact angle are known. This is straightforward, but the real problem is in the estimation of the solid-liquid interfacial energy and various mathematical models have been used to estimate it. These models fall into two categories: the surface tension component and the equation of state approaches. According to the first, γ_{sl} depends on γ_{la} and γ_{sa}, as well as the specific intermolecular interactions, i.e. components of the surface energy. The second, however says, γ_{sl} depends only on γ_{la} and γ_{sa} and a thermodynamic relation of the type $\gamma_{sl} = f(\gamma_{la}, \gamma_{sa})$ exists. A summary of these models is presented here, but details can be found in the reviews of Sharma and Hanumantha Rao (2002) and Della Volpe *et al.* (2004).

2.2.1 BERTHELOT'S GEOMETRIC MEAN COMBINING RULE AND ANTONOW'S RULE

The geometric mean combining rule is used for the determination of the interfacial tension from the surface tensions of the two phases. The way the dispersion energy coefficients C_6^{ij} are written in terms of C_6^{ii} and C_6^{jj} as $C_6^{ij} = (C_6^{ii}C_6^{jj})^{0.5}$, where i and j represent the phases, in the treatment of London theory of dispersion forces is used as a basis. Berthelot assumed that the interfacial work of adhesion W_{sl} is equal the geometric mean of the cohesion work W_{ss} of a solid and cohesion work W_{ll} of the liquid, i.e. $W_{sl} = (W_{ss}W_{ll})^{0.5}$, where $W_{ss} = 2\gamma_{sa}$ and $W_{ll} = 2\gamma_{la}$. However, according to the Dupre (1869) equation, $W_{sl} = \gamma_{sa} + \gamma_{la} - \gamma_{sl}$, and thus

$$\gamma_{sl} = \gamma_{sa} + \gamma_{la} - 2(\gamma_{sa}\gamma_{la})^{0.5} \text{ (Berthelot equation)} \tag{2.30}$$

According to Antonow, γ_{sl} can be expressed at once in terms of the γ_{sa} and γ_{la} as

$$\gamma_{sl} = |\gamma_{la} - \gamma_{sa}| \tag{2.31}$$

This is an empirical rule as there is no theoretical background to it (Kwok and Neumann 1999). The Berthelot and the Antonow rules are based on the equation of state approach.

2.2.2 BANGHAM AND RAZOUK MODEL

This model recognizes the importance of vapor adsorption on the solid surface and introduced a term to Young's equation known as the spreading pressure π_e so that $\gamma_{sa} = \gamma_s - \pi_e$, where $\gamma_s = \gamma_{sa} + \pi_e$. Based on this model, the Young's equation is

$$\gamma_s = \gamma_{sl} + \gamma_{la} \cos\theta + \pi_e \qquad (2.32)$$

2.2.3 THE ZISMAN APPROACH

Zisman introduced the concept of critical surface tension γ_c as an empirical means of determining the wettability of solid surfaces. The γ_c is obtained by plotting cos θ against the surface tensions of a series of liquids and finding the point at which the resulting curve (usually rectilinear) intercepts the line at $\cos\theta = 1$. This represents the critical surface tension, *i.e.* the tension that divides the liquids forming zero contact angle on the solid surface from those forming a contact angle greater than zero. Liquids with surface tension below the γ_c value spread on the solid surface. If the critical surface tension is considered to give an indication of the surface tension of the solid, then by using this method: (a) it is possible to obtain the total solid surface energy of an apolar solid by using series of homologous apolar liquids, *e.g.* *n*-alkanes, and (b) it is possible to obtain only the dispersion component of the total surface energy of a polar solid by using series of homologous apolar liquids like *n*-alkanes. Unfortunately, deviation from the rectilinear nature of Zisman plot is obtained when polar liquids are used on polar and apolar solids, making it impossible to determine any component of the solid surface energy. Also, it is possible to obtain different values of γ_c for a particular solid surface, depending on what liquid series is used in the experiment.

2.2.4 GOOD AND GIRIFALCO APPROACH

This approach introduces a constant ϕ (2.33) in the Berthelot's equation. This constant is specific to the solid-liquid system.

$$\gamma_{sl} = \gamma_{sa} + \gamma_{la} - 2\phi \ (\gamma_{sa}\gamma_{la})^{0.5} \qquad (2.33)$$

The

$$\phi = \frac{(\gamma_a + \gamma_b - \gamma_{ab})}{2\sqrt{\gamma_a\gamma_b}} \qquad (2.34)$$

For 'regular' systems, *i.e.* where the adhesive forces across the interface are of the same type, $\phi = 1$. When the predominant forces within the separate phases are of unlike types, *e.g.* London van der Waals forces and metallic bonding, $\phi < 1$. For cases where there are specific interactions between the molecules of the two phases, $\phi > 1$.

2.2.5 FOWKES AND ZETTLEMOYER APPROACHES

Fowkes pioneered the surface component approach. The model considers surface tension as a measure of the attractive force between the surface layer and liquid phase and that these forces and their contributions to the free energy are additive. Therefore, the surface tension of liquids as well as the surface energy of solids are considered to be composed of independent additive terms arising from London dispersion interactions γ^d which are predominant, polar interactions γ^p, hydrogen bonding γ^h, metallic bonding γ^m, and many others as given in Equation (2.35).

$$\gamma = \gamma^d + \gamma^p + \gamma^h + \gamma^m + ... \qquad (2.35)$$

London dispersion forces exist in all types of matter and always result in attractive force between adjacent atoms or molecules (no matter how dissimilar their nature may be). They result from the interaction of fluctuating electronic dipoles with the induced dipoles in neighbouring atoms or molecules. The effects of fluctuating dipoles cancel out, but not that of the induced dipoles. The dispersion forces contribute to the cohesion in all substances and are independent of other intermolecular forces. However, their magnitude depends on the type of material and density. The polar interactions include the Keesom and Debye forces.

Fowkes studied mainly two-phase systems containing a substance (solid or liquid) in which the dispersion interactions are only the operating forces and arrived at Equation (2.36),

$$\gamma_{sl} = \gamma_{sa} + \gamma_{la} - 2(\gamma_{sa}^d \gamma_{la}^d)^{0.5} \qquad (2.36)$$

which is clearly the Berthelot's equation in terms of interfacial London interactions. Equation (2.36) combines with the Young equation to give (2.37).

$$\cos\theta = 2\sqrt{\gamma_{sa}^d}\left(\frac{\gamma_{la}^d}{\gamma_{la}}\right) - \frac{\pi_e}{\gamma_{la}} - 1 \qquad (2.37)$$

If the spreading pressure term is ignored, then a plot of $\cos\theta$ against $\gamma_{la}^d/\gamma_{la}$ gives a straight line with the origin at $\cos\theta = -1$ and gradient $2\sqrt{\gamma_{sa}^d}$. Because the origin is fixed, one contact angle data is enough to determine the dispersion force component of the solid surface energy.

The spreading pressure term is assumed to be zero only for the system where high energy liquids are brought in contact with low energy solids. This is because all theoretical and experimental evidence predicts that adsorption of high-energy materials cannot reduce the surface energy of a low energy material, *e.g.* adsorbing water does not reduce the surface tension of a liquid hydrocarbon. Finally, a liquid with a non-zero contact angle on a given low energy solid asserts that the liquid possesses a higher energy and the spreading pressure is zero. However, this is true only for solids interacting *via* dispersion forces only. For high-energy solids like metals and graphite, which water does not wet (*i.e.* the contact angle is more than 0°), but rather adsorbs, the spreading pressure is quite appreciable.

Rather than using geometric mean to combine the dispersion component of solid and liquid, Zettlemoyer used arithmetic mean and arrived at Equation (2.38). However, this approach does not have a significant scientific basis.

$$\gamma_{sl} = \gamma_{sa} + \gamma_{la} - (\gamma_{sa}^d + \gamma_{la}^d) \tag{2.38}$$

2.2.6   OWENS AND WENDT, AND WU APPROACHES

Owens and Wendt (1969) significantly changed the Fowkes idea by assuming that all the components occurring on the right-hand side of Equation (2.35), except γ^d, are associated with the polar interaction γ^p. This resulted to Equation (2.39).

$$\gamma_{sl} = \gamma_{sa} + \gamma_{la} - 2(\gamma_{sa}^d \gamma_{la}^d)^{0.5} - 2(\gamma_{sa}^p \gamma_{la}^p)^{0.5} \tag{2.39}$$

Because the polar interaction defined by Fowkes differs from that by Owens and Wendt, the meaning of γ^p in Equations (2.35) and (2.39) are different.

Wu accepted the Owens and Wendt approach of dividing the surface energy into two parts and used harmonic, rather than geometric, mean to combine the polar and dispersion components of the solid and liquid surface energies to obtain Equation (2.40).

$$\gamma_{sl} = \gamma_{sa} + \gamma_{la} - 4\left[\left(\frac{\gamma_{sa}^d \gamma_{la}^d}{\gamma_{sa}^d + \gamma_{la}^d}\right) + \left(\frac{\gamma_{sa}^p \gamma_{la}^p}{\gamma_{sa}^p + \gamma_{la}^p}\right)\right] \tag{2.40}$$

2.2.7   WARD AND NEUMANN APPROACH

Ward and Neumann gave thermodynamic proof for the existence of the equation of state. Neumann showed, in a series of papers (Neumann *et al.* 1974, Li and Neumann 1990, 1992, and Kwok and Neumann 1999) that

$$\gamma_{sl} = \frac{\sqrt{\gamma_{sa}} - \sqrt{\gamma_{la}}}{1 - 0.015\sqrt{\gamma_{sa}\gamma_{la}}} \tag{2.41}$$

$$\gamma_{sl} = \gamma_{sa} + \gamma_{la} - 2\sqrt{\gamma_{sa}\gamma_{la}}\exp[-\beta_1(\gamma_{la} - \gamma_{sa})^2] \tag{2.42}$$

$$\gamma_{sl} = \gamma_{sa} + \gamma_{la} - (2\sqrt{\gamma_{sa}\gamma_{la}})[1 - \beta_2(\gamma_{la} - \gamma_{sa})^2] \tag{2.43}$$

The coefficients β_1 and β_2, as determined experimentally, are 0.0001247 and 0.0001057, respectively.

2.2.8   VAN OSS-CHAUDHURY-GOOD APPROACH

This approach divides the γ_{sa} into two parts. The first part contains long-range interactions, *i.e.* London, Keesom and Debye, known as the Lifshitz-van der Waals component γ^{LW}. The second contains short-ranged interactions (acid-base) known as the acid-base component γ^{AB}. The latter component is considered equal $2(\gamma^+\gamma^-)^{0.5}$, where γ^+ and γ^- are the acidic and basic constituents, respectively, associated with the acid-base interactions. In other words, the polar component of surface energy is further divided into electron-accepting γ^+ and electron-donating γ^- parameters. According to this approach, the γ_{sl} is given as

$$\gamma_{sl} = (\sqrt{\gamma_{sa}^{LW}} - \sqrt{\gamma_{la}^{LW}})^2 + 2[(\sqrt{\gamma_{sa}^+} - \sqrt{\gamma_{la}^+})^2 \cdot (\sqrt{\gamma_{sa}^-} - \sqrt{\gamma_{la}^-})^2] \qquad (2.44)$$

2.3 WETTING AND NON-WETTING SURFACES AND BIOINSPIRED NON-WETTING SURFACES

2.3.1 WETTING SURFACES

With water drops, certain solid surfaces exhibit low equilibrium apparent contact angle values below 90° with an associated high hysteresis. Such solid surfaces are said to be 'hydrophilic' because of their high affinity for water molecules. Hydrophilic surfaces contain polar groups like $-OH$, $-NH_2$, $-COOH$, $-OSO_3H$, $-NH_3^+$, $-COO^-$ and $-OSO_3^-$. Similarly, surfaces exhibiting this characteristic with oil drops are said to be 'oleophilic' and may contain polar or nonpolar groups. Surfaces that are hydrophilic as well as oleophilic are said to be 'amphiphilic'. These surfaces have a relatively high surface energy. Wetting surfaces are sometimes modified *via* surface chemical modification to obtain non-wetting ones.

2.3.2 NON-WETTING AND BIOINSPIRED NON-WETTING SURFACES

A non-wetting state is witnessed when a liquid drop sits almost spherically on the surfaces of a solid with an equilibrium contact angle value above 90° with relatively low hysteresis. The surface is said to be 'hydrophobic' when the liquid is water and 'oleophobic' when the liquid is oil. Surfaces that are both hydrophobic and oleophobic are said to be 'omniphobic'. 'Superhydrophobic' and 'superoleophobic' surfaces are those in which the equilibrium contact angle made by the water and oil drops, respectively, is $\gg 90°$ (usually $\geq 150°$). Similarly, surfaces that are both superhydrophobic and superoleophobic are said to be 'superomniphobic'. These surfaces have a relatively low surface energy. Their creation by scientists has been inspired by nature (biological species), plants, and animals alike, whose surfaces remain largely non-wetting when in contact with liquids and hence the term 'bioinspired non-wetting' surfaces. The surfaces of the lotus (*Nelumbo nucifera*) leaf and butterfly (*Parnassius glacialis*) wings, Figure 2.8, which are completely non-wetted by water drops, are typical examples of these surfaces.

Roughness is a major characteristic of these surfaces as evidenced in Figure 2.8 and the creation of the so-called biomimetic or bioinspired surfaces is based mainly on this. In addition to being rough, these surfaces often contain non-polar groups like alkyl groups, silicones $-[Si(CH_3)_2O]-$ and fluorocarbons $-C_nF_{2n+1}$ (Chhatre *et al.* 2010). This is one reason why these surfaces exhibit low surface energy. For example, Tuteja *et al.* (2007) have used re-entrant surface curvature (*i.e.* convex topographies), in addition to chemical composition and roughness, to design

FIGURE 2.8   (a) Photograph of a water drop on the surfaces of the Lotus leaf. (From Bhushan, B. *et al., Phil. Trans. R. Soc. A*, 367, 1631–1672, 2009.) (b) SEM image of the Lotus leaf showing its microstructural nature. (From Koch, K. *et al., Prog. Mater. Sci.*, 54, 137–178, 2009.) (c) Photograph of a transparent butterfly (*Parnassius glacialis*) and a water drop on one of its wings and (d) SEM image of *Parnassius glacialis*' wing showing its rough microstructures. (From Goodwyn, P.P. *et al., Naturwissenschaften*, 96, 781–787, 2009.)

superoleophobic surfaces that were not wetted by many oils including alkanes (C_8–C_{16}). Because these surfaces are rough, liquid drops on them are either in the Wenzel or Cassie-Baxter wetting state. Non-wetting surfaces have several applications. They are the basis for the creation of self-cleaning surfaces, anti-fog windows, and drag-reduction systems. They are also an efficient way for transportation of liquids. In medicine, their applications range from high-throughput cell culture platforms to biomedical devices.

Powdered particles (nm–μm) are also hydrophilic, hydrophobic, superhydrophobic, oleophilic, oleophobic, superoleophobic, or superomniphobic, depending on whether there are wetted by a bulk liquid phase or not (Tyowua 2014). Unlike in the case of solid surfaces, powdered particles are said to be hydrophilic if they are completely wetted by water and go into it upon placing them on the surfaces of a bulk water phase. Hydrophobic particles, however, are only partially wetted by water and remain on its surfaces once in contact with it. As illustrated in Figure 2.9 for a spherical powdered particle, the three-phase contact angle is 0° in the former and >90° in the latter. It is worthy to note that $0° < \theta < 90°$ for some hydrophilic particles. Superhydrophobic particles are also partially wetted by water and remain on the surfaces of water once in contact with it, but the three-phase contact angle is >>90° (usually ≥ 150°). Similarly, oleophilic particles are completely wetted by oils and go into them while oleophobic particles are partially wetted by oils and

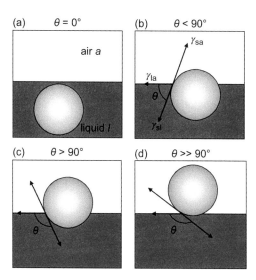

FIGURE 2.9   Illustration of a spherical powdered particle at a liquid-air interface when the three-phase contact angle θ is (a) 0° (*'philic*), (b) θ < 90° (*'philic*), (c) θ > 90° (*'phobic*) and (d) θ >> 90° (*super'phobic*). (From Binks, B.P. and Horozov, T.S., *Colloidal Particles at Liquid Interfaces*, Cambridge University Press, Cambridge, 2006.)

remain on their surfaces once in contact with them. The corresponding three-phase contact angles for oleophilic and oleophobic particles are 0° and >90°, respectively. Superoleophobic particles too are partially wetted by oils and remain on oil surfaces once in contact with them, and the three-phase contact angle is also >>90°. Particles that are both hydrophobic and oleophobic are said to be omniphobic while those that are both superhydrophobic and superoleophobic are said to be superomniphobic. It is important to remember that because it is difficult to measure the three-phase contact angle powdered particles make with bulk liquid phases; these angles are measured on solid substrates composed of the particles or on pellets of the particles. Because these angles are not the true three-phase contact angles, they are known as apparent contact angles. Due to the low surface tension of many oils, it should be noted that all oleophobic and superoleophobic particles are hydrophobic and superhydrophobic, respectively, but not all hydrophobic and superhydrophobic particles are oleophobic and superoleophobic, as the surface tension of water is relatively high. Finally, it should be noted that liquid marble preparation requires powdered particles to be 'phobic'.

2.4   CONCLUSION

The wetting of solid surfaces and powdered particles is discussed in terms of the equilibrium particle-liquid-air three-phase contact angle θ, made by a liquid drop once on the surfaces of the solid substrate and the powdered particle once on the surfaces of a bulk liquid phase, respectively. This is continued with the various methods of measuring θ for solid surfaces and powdered particles, a brief discussion of contact angle hysteresis, and the different wetting states. Thereafter, the surface energy (akin to liquid surface tension) of solid surfaces and powdered particles is described in detail, including its measurement. Finally, using the foregoing knowledge, solid surfaces and powdered particles are described as hydrophilic, oleophilic, hydrophobic, oleophobic, omniphobic, superhydrophobic, superoleophobic, or super-omniphobic, pointing out that the phobic particles are especially important in liquid marble formation.

EXERCISES

DISCUSSION QUESTIONS

Question 1
a. What is three-phase contact angle?
b. What is wetting? State the three wetting states you know.
c. What is spreading coefficient?

Question 2
a. Describe the various ways through which contact angles can be measured.
b. Differentiate between the sessile drop and captive bubble methods of contact angle measurement.
c. Differentiate between the Wilhelmy balance and capillary rise at a vertical plate methods of contact angle measurement.

Question 3
a. What is contact angle hysteresis?
b. Highlight the usefulness and uselessness of contact angle hysteresis.
c. Differentiate between the Young, the Wenzel, and the Cassie-Baxter wetting states.

Question 4
a. What is the surface energy?
b. Discuss the various means of measuring the surface energy of solid surfaces.
c. Differentiate between non-wetting and bioinspired non-wetting surfaces.

Question 5
a. Explain the following terms in relation to solid surfaces and powdered particles:
(i) hydrophilic, (ii) oleophilic, (iii) hydrophobic, (iv) oleophobic, (v) omniphobic, (vi) superhydrophobic, (vii) superoleophobic, and (viii) superomniphobic.

NUMERICAL QUESTIONS

Question 1
a. Derive Young's equation.
b. Taking the density and surface tension of water at 25°C as 997 kg m^{-3} and 72 mN m^{-1}, respectively, calculate the contact angle of water in a Wilhemy balance experiment if the perimeter of the plate is 48 mm

and the force change is 3.5 mN, while the volume of water displaced is 1.5×10^{-4} cm^3 and the density of air at the prevailing temperature is 1.184 kg m^{-3}.

c. A rise height of 2.2×10^{-4} cm was observed when a vertical glass plate was immersed in water at 25°C. If the density and surface tension of the water are 997 kg m^{-3} and 72 mN m^{-1}, respectively, while the density of air at the prevailing condition is 1.184 kg m^{-3}, calculate the contact angle of the glass plate with the water.

Question 2

a. Calculate the contact angle the liquid drops shown in Figure 2.10 make with the solid surfaces using Equation (2.16). Compare the angles with the experimentally determined ones shown on the drop images and comment on your observation. Each scale bar is 0.5 mm.

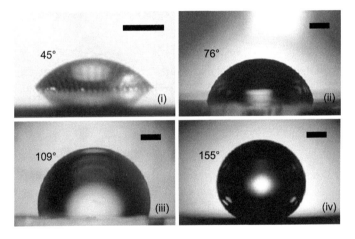

FIGURE 2.10 Images of liquid drops, (i) methanol, (ii) water, (iii) hexadecane in water and (iv) water, on solid surfaces at ambient conditions. (From Tuteja, A. *et al. PNAS*, 105, 18200–18205, 2008, and Jung, Y.C. and Bhushan, B., *Langmuir* 24, 14165–14173, 2009.)

b. The advancing and receding angles of 10 mg of a rain drop ($\gamma_{la} = 70$ mN m^{-1}, radius 0.5 mm) on a vertical window pane are 74 and 54°, respectively. Will the rain drop roll off or remain stuck to the pane?

Question 3

The advancing contact angle values of compressed disks of fluorinated sericite PF-8 particles with liquids of different surface tension values are shown in Table 2.1. Construct a Zisman plot and determine the critical surface tension of the surface if possible.

TABLE 2.1

Advancing Contact Angle θ Values of Drops (10 μL) of Liquids on Compressed Disks of Fluorinated Sericite PF-8 Particles in Air at 25°C

Liquid	Tension/ mN m^{-1}	θ/°
Water	72.8	145
Glycerol	64.0	137
Formamide	58.0	129
Ethylene glycol	48.0	129
Diiodomethane	50.8	126
α-Bromonaphthalene	44.4	120
n-Hexadecane	27.8	112

FURTHER READING

't Mannetje, D.J.C.M. Drops, Contact Lines and Electrowetting. Dietrich J.C.M. 't Mannetje, *Physics of Complex Fluids*. Enschede, the Netherlands: University of Twente, 2013.

Bormashenko, E.Y. *Wetting of Real Surfaces*. Berlin, Germany: Walter de Gruyter GmbH, 2013.

Della Volpe, C. and S. Siboni. "Use, Abuse, Misuse and Proper Use of Contact Angles: A Critical Review". *Rev. Adhesion Adhesives*. 3 (2015): 365–85.

Etzler, F.M. "Characterization of Surface Free Energies and Surface Chemistry of Solids". In *Contact Angle, Wettability and Adhesion*, Vol. 3, K.L. Mittal (Ed.). Boca Raton, FL: CRC Press, 2003.

Etzler, F.M. "Determination of the Surface Free Energy of Solids: A Critical Review". *Rev. Adhesion Adhesives*. 1 (2013): 3–45.

Quéré, D. "Rough Ideas on Wetting." *Physica A*. 313 (2002): 32–46.

Quéré, D. "Wetting and Roughness." *Annu. Rev. Mater. Res*. 38 (1) (2008): 71–99.

Sas, I., R.E. Gorga, J.A. Joines and K.A. Thoney. "Literature Review on Superhydrophobic Self-Cleaning Surfaces Produced by Electrospinning." *J. Polymer. Sci. B Polymer. Phy*. 50 (2012): 824–45.

REFERENCES

't Mannetje, D.J.C.M., C.U. Murade, D. van den Ende and F. Mugele. "Electrically Assisted Drop Sliding on Inclined Planes." *Appl. Phys. Lett*. 98 (1) (2011): 014102.

Bartell, F.E. and H.J. Osterhof. "Determination of the Wettability of a Solid by a Liquid." *Ind. Eng. Chem*. 19 (11) (1927): 1277–80.

Bezuglyi, B.A., O.A. Tarasov and A.A. Fedorets. "Modified Tilting-Plate Method for Measuring Contact Angles." *Colloid J*. 63 (6) (2001): 668.

Bhushan, B., Y.C. Jung and K. Koch. "Micro-, Nano- and Hierarchical Structures for Superhydrophobicity, Self-Cleaning and Low Adhesion." *Phil. Trans. R. Soc. A* 367 (2009): 1631–72.

Bigelow, W.C., D.L. Pickett and W.A. Zisman. "Oleophobic Monolayers I. Films Adsorbed from Solution in Non-Polar Liquids." *J. Colloid Sci*. 6 (6) (1964): 513–38.

Binks, B.P., T. Sekine and A.T. Tyowua. "Dry Oil Powders and Oil Foams Stabilized by Fluorinated Clay Platelet Particles." *Soft Matter* 10 (4) (2014): 578–89.

Binks, B.P. and Horozov, T.S. *Colloidal Particles at Liquid Interfaces*. Cambridge, UK: Cambridge University Press, 2006.

Bormashenko, E. "Wetting Transitions on Biomimetic Surfaces." *Phil. Trans. R. Soc. A* 368 (2010): 4695–711.

Budziak, C.J. and A.W. Neumann. "Automation of the Capillary Rise Technique for Measuring Contact Angles." *Colloid Surf*. 43 (2) (1990): 279–93.

Cassie, A.B.D. "Contact Angles." *Discuss. Faraday Soc*. 3 (1948): 11–16.

Cassie, A.B.D. and S. Baxter. "Wettability of Porous Surfaces." *Trans. Faraday Soc*. 40 (1944): 546–51.

Chaudhury, M.K. "Interfacial Interaction between Low-Energy Surfaces." *Mater. Sci. Eng*. R16 (1996): 97–159.

Cheng, P., D. Li, L. Boruvka and Y. Rotenberg. "Automation of Axisymmetric Drop Shape Analysis for Measurements of Interfacial Tensions and Contact Angles." *Colloid Surf*. 43 (1990): 151.

Chhatre, S.S., J.O. Guardado, B.M. Moore *et al*. "Fluoroalkylate Silicon-Containing Surfaces-Estimation of Solid Surface Energy." *ACS Appl. Mater. Interfaces* 2 (12) (2010): 3544–54.

Clint, J.H. and A.C. Wicks. "Adhesion under Water: Surface Energy Considerations." *Int. J. Adhes. Adhes*. 21 (4) (2001): 267–73.

del Rio, O.I. and A.W. Neumann. "Axisymmetric Drop Shape Analysis: Computational Methods for the Measurement of Interfacial Properties from the Shape and Dimensions of Pendant and Sessile Drops." *J. Colloid Interface Sci.* 196 (1997): 136.

Della Volpe, C., D. Maniglio, M. Brugnara, S. Siboni and M. Morra. "The Solid Surface Free Energy Calculation: I. In Defence of the Multicomponent Approach." *J. Colloid Interface Sci.* 271 (2) (2004): 434–53.

Diggins, D., L.G.J. Fokkink and J. Ralston. "The Wetting of Angular Quartz Particles: Capillary Pressure and Contact Angles." *Colloid Surf.* 44 (1990): 299.

Dunstan, D. and L.R. White. "A Capillary Pressure Method for Measurement of Contact Angles in Powders and Porous Media." *J. Colloid Interface Sci.* 111 (1986): 60.

Dupre, A. *Theorie Mecanique De La Chaleur.* Paris: Gauthier-Villan, 1869.

Eral, H.B., D.J.C.M.'t Mannetje and J.M. Oh. "Contact Angle Hysteresis: A Review of Fundamentals and Applications." *Colloid Polym. Sci.* 291 (2) (2012): 247–60.

Extrand, C.W. and Y. Kumagai. "An Experimental Study of Contact Angle Hysteresis." *J. Colloid Interface Sci.* 191 (2) (1997): 378–83.

Extrand, C.W. and Y. Kumagai. "Contact Angle and Hysteresis on Soft Surfaces." *J. Colloid Interface Sci.* 184 (1) (1996): 191–200.

Fowkes, F.M. and W.D. Harkins. "The State of Monolayers Adsorbed at the Interface Solid-Aqueous Solution." *J. Am. Chem. Soc.* 62 (12) (1940): 3377–86.

Fox, H.W. and W.A. Zisman. "The Spreading of Liquids on Low Energy Surfaces. I. Polytetrafluoroethylene." *J. Colloid Sci.* 5 (6) (1950): 514–31.

Goodwyn, P.P., Y. Maezono, N. Hosoda and K. Fujisaki. "Waterproof and Translucent Wings at the Same Time: Problems and Solutions in Butterflies." *Naturwissenschaften* 96 (2009): 781–87.

Harris, D.J., H. Hu, J.C. Conrad and J.A. Lewis. "Patterning Colloidal Films *via* Evaporative Lithography." *Phys. Rev. Lett.* 98 (14) (2007): 148301.

Hu, D.L. and J.W.M. Bush. "The Hydrodynamics of Water-Walking Arthropods." *J. Fluid Mech.* 644 (2010): 5–33.

Jordan, D.O. and J.E. Lane. "A Thermodynamic Discussion of the Use of a Vertical-Plate Balance for the Measurement of Surface Tension." *Austral. J. Chem.* 17 (1964): 7.

Jung, Y.C. and B. Bhushan. "Wetting Behaviour of Water and Oil Droplets in Three-Phase Interfaces for Hydrophobicity/Philicity and Oleophobicity/Philicity." *Langmuir* 24 (2009): 14165–73.

Kabza, K.G., J.E. Gestwicki and J.L.J. McGrath. "Contact Angle Goniometry as a Tool for Surface Tension Measurements of Solids, Using Zisman Plot Method. A Physical Chemistry Experiment." *J. Chem. Educ.* 77 (2000): 63.

Kalantarian, A., R. David and A.W. Neumann. "Methodology for High Accuracy Contact Angle Measurement." *Langmuir* 25 (24) (2009): 14146–54.

Koch, K., B. Bhushan and W. Barthlott. "Multifunctional Surface Structures of Plants: An Inspiration for Biomimetics." *Prog. Mater. Sci.* 54 (2009): 137–78.

Krasovitski, B. and A. Marmur. "Drop Down the Hill: Theoretical Study of Limiting Contact Angles and the Hysteresis Range on a Tilted Plate." *Langmuir* 21 (9) (2005): 3881–85.

Kwok, D.Y. and A.W. Neumann. "Contact Angle Measurement and Contact Angle Interpretation." *Adv. Colloid Interface Sci.* 81 (1999): 167–249.

Kwok, D.Y., C.J. Budziak and A.W. Neumann. "Measurements of Static and Low Rate Dynamic Contact Angles by Means of an Automated Capillary Rise Technique." *J. Colloid Interface Sci.* 173 (1) (1995): 143–50.

Kwok, D.Y., R. Lin and A.W. Neumann. "Low-Rate Dynamic and Static Contact Angles and the Determination of Solid Surface Tensions." *Colloid Surf. A* 116 (1) (1996): 63–77.

Li, D. and A.W. Neumann. "A Reformation of the Equation of State for Interfacial Tensions." *J. Colloid Interface Sci.* 137 (1990): 304–7.

Li, D. and A.W. Neumann. "Equation of State for Interfacial Tensions of Solid-Liquid Systems." *Adv. Colloid Interface Sci.* 39 (1992): 299–345.

Li, F. and F. Mugele. "How to Make Sticky Surfaces Slippery: Contact Angle Hysteresis in Electrowetting with Alternating Voltage." *Appl. Phys. Lett.* 92 (24) (2008): 2441081–83.

Macdougall, G. and C. Ockrent. "Surface Energy Relations in Liquid/Solid Systems. I. The Adhesion of Liquids to Solids and a New Method of Determining the Surface Tension of Liquids." *Proc. R. Soc. A* 180 (981) (1942): 151–73.

Neumann, A.W. and R.J. Good. "Experimental Methods." In *Surface and Colloid Science*, R.J. Good and R.R. Stromberg (Ed.). Plenum Publishing: New York, 1979.

Neumann, A.W., R.J. Good, C.J. Hope and M. Sejpal. "An Equation-of-State Approach to Determine Surface Tensions of Low-Energy Solids from Contact Angles." *J. Colloid Interface Sci.* 49 (1974): 291–304.

Owens, D.K. and R.C. Wendt. "Estimation of the Surface Free Energy of Polymers." *J. Appl. Polym. Sci.* 13 (1969): 1741–47.

Pierce, E., F.J. Carmona and A. Amirfazil. "Understanding of Sliding and Contact Angle Results in Tilted Plate Experiments." *Colloid Surf. A* 323 (1) (2008): 73–82.

Princen, H.M. "Capillary Phenomena in Assemblies of Parallel Cylinders I. Capillary Rise between Two Cylinders." *J. Colloid Interface Sci.* 30 (1) (1969): 69–75.

Ragil, K., D. Bonn, D. Broseta and J. Meunier. "Wetting of Alkanes on Water from a Cahn-Type Theory." *J. Chem. Phys.* 105 (1996): 5160.

Rotenberg, Y., L. Boruvka and A.W. Neumann. "Determination of Surface Tension and Contact Angle from the Shapes of Axisymmetric Fluid Interfaces." *J. Colloid Interface Sci.* 93 (1983): 169.

Sharma, P.K. and K. Hanumantha Rao. "Analysis of Different Approaches for Evaluation of Surface Energy of Microbial Cells by Contact Angle Goniometry." *Adv. Colloid Interface Sci.* 98 (3) (2002): 341–463.

Smithwich, R.W. "Contact-Angle Studies of Microscopic Mercury Droplets on Glass." *J. Colloid Interface Sci.* 123 (2) (1988): 482–85.

Spelt, J.K., Y. Rotenberg, D.R. Absolom and A.W. Neumann. "Sessile-Drop Contact Angle Measurements Using Axisymmetric Drop Shape Analysis." *Colloids Surf.* 24 (1987): 127.

Taggart, A.F., T.C. Taylor and C.R. Ince. "Experiments with Flotation Agents." *Am. Inst. Min. Metall. Pet. Eng.* 87 (1930): 285–386.

Tuteja, A., W. Choi, J.M. Mabry, G.H. McKinley and R.E. Cohen. "Robust Omniphobic Surfaces." *PNAS* 105 (2008): 18200–205.

Tuteja, A.C., W. Choi, M. Ma et al. "Designing Superoleophobic Surfaces." *Science* 318 (5856) (2007): 1618–22.

Tyowua, A.T. "Solid Particles at Fluid Interfaces: Emulsions, Liquid Marbles, Dry Oil Powders and Oil Foams." PhD thesis of the University of Hull, UK, 2014.

Tyowua, A.T., S.G. Yiase and B.P. Binks. "Double Oil-in-Oil-in-Oil Emulsions Stabilized Solely by Particles." *J. Colloid Interface Sci.* 488 (2017): 127–34.

Washburn, E.W. "The Dynamics of Capillary Flow." *Phys. Rev.* 17 (1921): 273.

White, L.R. "Capillary Rise in Powders." *J. Colloid Interface Sci.* 90 (1982): 536–38.

Whyman, G., E. Bormashenko and T. Stein. "The Rigorous Derivation of Young, Cassie-Baxter and Wenzel Equations and the Analysis of the Contact Angle Hysteresis Phenomenon." *Chem. Phys. Lett.* 450 (2008): 355–59.

Yuan, Y. and R.T. Lee. "Contact Angle and Wetting Properties." In *Surface Science Techniques*, G. Bracco and B. Holst (Eds.). Berlin Heidelberg: Springer-Verlag, 2013.

Zhang, W., M. Wahlgren and B. Sivik. "Membrane Characterization by the Contact Angle Technique II. Characterization of UF-Membrane and Comparison between the Captive Bubble and Sessile Drop as Methods to Obtain Water Contact Angles." *Desalination* 72 (3) (1989): 263–73.

3 Sticking and Non-sticking Drops

3.1 SHAPE OF DROPS

The equilibrium contact angle of a liquid drop resting on an ideal solid surface is a local quantity and independent of the size of the liquid drop, provided intermolecular forces and line tension effects are absent. However, the shape of a liquid drop resting on a solid surface depends on its size so that relatively small ones are quasi-spherical (*i.e.* they make spherical cap), while relatively large ones are flattened by gravity and are puddle-shaped. These drop shapes are governed by the interplay of surface tension forces and the forces of gravity. The associated surface energy scales as $\gamma_{la}R_o^2$ while the associated gravitational energy scales as $\rho g R_o^4$. The γ_{la} is the liquid surface tension, R_o is the drop radius, ρ is the liquid density, and g is the acceleration of gravity. Generally, there exists a certain length, known as the capillary length κ^{-1}, beyond which the effect of gravity becomes significant on a liquid drop at rest. By comparing the Laplace pressure γ/κ^{-1} to the hydrostatic pressure $\rho g \kappa^{-1}$ at a depth κ^{-1} in a liquid of density ρ submitted to earth's gravity g, one can obtain κ^{-1} upon equating the two pressures as

$$\kappa^{-1} = \left(\frac{\gamma_{la}}{\rho g} \right)^{0.5} \tag{3.1}$$

Suppose that R_o is the initial radius of the spherical drop before deposition on the solid surface, quasi-spherical liquid drops are obtained when $R_o \ll \kappa^{-1}$ while puddle-shaped ones are obtained when $R_o \gg \kappa^{-1}$. The surface tension forces dominate in the former while the gravity forces dominate in the latter.

Example 3.1: Calculating the Capillary Length of a Liquid Drop

Calculate the capillary length of water maintained at 25°C where the surface tension is 72 mN m^{-1} and density is 997 kg m^{-3}. Based on the capillary length, would a 40 μL water drop at the same temperature be quasi-spherical or puddle-shaped.

Method

Convert the surface tension value from mN m^{-1} to N m^{-1} and substitute into Equation (3.1) along with the density value as well as the gravitational acceleration value (9.8 m s^{-2}) and solve for κ^{-1} at once. Compare the radius R_o of the drop with κ^{-1}, bearing in mind that $R_o \ll \kappa^{-1}$ for quasi-spherical drops and $R_o \gg \kappa^{-1}$ for puddles.

Answer

$$\kappa^{-1} = \left(\frac{\gamma_{la}}{\rho g}\right)^{0.5} = \left(\frac{72 \times 10^{-3} \text{ N m}^{-1}}{997 \text{ kg m}^{-3} \times 9.8 \text{ m s}^{-2}}\right)^{0.5} = 2.7 \times 10^{-3} \text{ m} = 2.7 \text{ mm}$$

The volume Ω of a sphere is given as

$$\Omega = \frac{4}{3}\pi R_o^3$$

$$\Rightarrow R_o = \left(\frac{3\Omega}{4\pi}\right)^{\frac{1}{3}} = \left(\frac{3 \times 40 \times 10^{-9} \text{ m}^3}{4\pi}\right)^{\frac{1}{3}} = 2.1 \times 10^{-3} \text{ m} = 2.1 \text{ mm}$$

Therefore, because R_o is less than κ^{-1}, the drop will be quasi-spherical.

For quasi-spherical drops, spherical caps, the radius δ (Figure 3.1) of the solid-liquid contact zone is

$$\delta = R \sin\theta \tag{3.2}$$

where R is the radius of curvature of the drop. The drop volume Ω, Equation (3.3), is equivalent to the volume Ω' of the spherical cap, Equation (3.4).

$$\Omega = \frac{4}{3}\pi R_o^3 \tag{3.3}$$

$$\Omega' = \frac{\pi}{3}(2 + \cos\theta)(1 - \cos\theta)^2 R^3, \text{ where } R = \delta/\sin\theta \tag{3.4}$$

By equating Equations (3.3) and (3.4), one can obtain δ as

$$\delta = 4^{1/3} \frac{\sin\theta}{(2 + \cos\theta)^{1/3}(1 - \cos\theta)^{2/3}} R_o \tag{3.5}$$

The contact slowly diverges as the drop slowly wets the substrate (*i.e.* $\theta \to 0°$) and vanishes as $\pi-\theta$ approaches zero (*i.e.* as the drop tends to dry). For puddles, a region

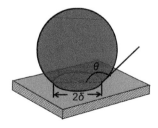

FIGURE 3.1 Schematic of a liquid drop resting on a planar solid substrate with a contact zone δ and an equilibrium three-phase contact angle of θ.

(size ~ κ^{-1}) close to the contact zone remains curved by the existence of a non-zero contact angle and their thickness H can be calculated by using Equation (3.6) (Taylor and Michael 1973).

$$H = 2\kappa^{-1}\sin\left(\frac{\theta}{2}\right) \tag{3.6}$$

The thickness is a monotonous function of θ and for very large drops it increases from 0 to $2\kappa^{-1}$ as θ increases from 0° to 180°. For drops of intermediate sizes, the radial curvature of the drop induces a Laplace overpressure which increases the height of the puddle slightly. In the non-wetting state ($\theta = 180°$), H is a function of R_0 and passes a maximum of about $2.1\kappa^{-1}$ for R_0 of the order of $3.2\kappa^{-1}$ before decreasing towards the asymptotic value $2\kappa^{-1}$ (Aussillous and Quéré 2006).

Example 3.2: Describing How the Drop Thickness Varies with Contact Angle

For water puddles (maintained at 25°C where the surface tension is 72 mN m^{-1} and density is 997 kg m^{-3}) on a hydrophobic planar solid substrate, describe how the thickness H varies with the contact angle θ in the range of 0° to 180°.

Method

Draw a table for H versus θ in the range of 0° to 180° using Equation (3.6). Draw a graph of H versus θ using the table and describe the behavior of the graph which represents the behavior of the puddles.

Answer

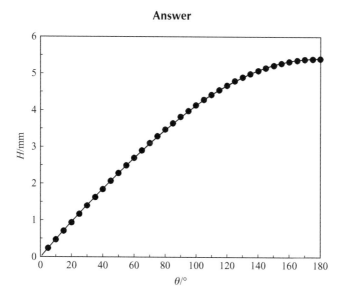

The puddle thickness increases linearly and gently before reaching the asymptotic value of $2\kappa^{-1}$ at 180°. Near the asymptotic value, the puddle thickness is independent of the contact angle and the graph plateaus as a result.

3.2   STICKING DROPS

The sticking of rain drops on glass windows, car windscreens, and many inclined surfaces illustrate that liquid drops can remain stuck to a surface. Consider the case of a liquid drop on a solid surface inclined at an angle α, Figure 3.2 (upper image),

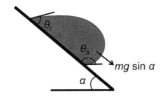

FIGURE 3.2   The upper part of the image shows the schematic of a small liquid drop resting on a plane inclined at an angle α to the horizontal. The drop makes a contact angle θ_r at the rear and θ_a at the front. The weight $mg \sin \alpha$ of the drop acts at the front parallel to the plane. The lower part of the image shows a plot of $\Delta\theta$ versus θ, in line with Equation (3.10), for a water drop (~ 0.9 µL, R_0 = 0.6 mm) at 25°C placed on a plane inclined at $\alpha = 80°$ to the horizontal.

where the contact angle of the drop at the rear is θ_r (receding angle) while that at the front is θ_a (advancing angle). In addition, let the hysteresis $\Delta\theta$ be small (*i.e.* $\Delta\theta \ll \theta_a$) so that the drop retains its shape of a spherical cap, bounded by a circular contact line. The drop would necessarily run down if the equilibrium contact angles were the same (*i.e.* $\theta_r \approx \theta_a$). The force that holds the drop in place comes from the asymmetric nature of the drop and the drop remains stuck if:

$$\pi\gamma_{la}\delta(\cos\theta_r - \cos\theta_a) \geq \rho g\Omega\sin\alpha \tag{3.7}$$

$$\Rightarrow \pi\gamma_{la}\delta(\cos\theta_r - \cos\theta_a) \geq mg\sin\alpha \tag{3.8}$$

By introducing an average angle θ,

$$\theta_r = \theta - \frac{\Delta\theta}{2} \text{ and } \theta_a = \theta + \frac{\Delta\theta}{2} \tag{3.9}$$

and expanding Equation (3.8) for small $\Delta\theta$, the sticking condition is obtained as

$$\Delta\theta \geq \Delta\theta_c = \frac{4^{2/3}}{3}(R_0\kappa)^2 \frac{(2+\cos\theta)^{1/3}(1-\cos\theta)^{2/3}}{\sin^2\theta}\sin\alpha \tag{3.10}$$

For relatively small drops ($R_0 \ll \kappa^{-1}$) on moderately inclined substrates (*i.e.* small α), Equation (3.10) defines a contact angle hysteresis which remains small (even at the threshold) as considered above. On the contrary, relatively large drops where $R_0 \gg \kappa^{-1}$ move down as gravity dominates over the surface forces responsible for sticking the drop.

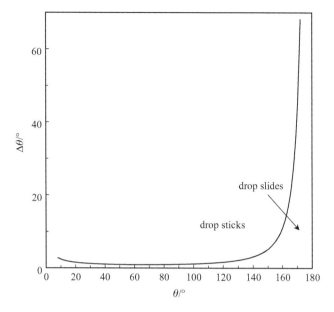

Figure 3.2 (lower part of the image) shows a plot of $\Delta\theta$ versus θ, in line with Equation (3.10), for a water drop (~0.9 μL, $R_0 = 0.6$ mm) at 25°C placed on a plane inclined at 80° to the horizontal. The figure is divided into two regions by the curve $\Delta\theta_c$: that in which $\Delta\theta$ is relatively large (or θ is relatively small) and the drop sticks and that in which $\Delta\theta$ is relatively small (or θ is relatively large) and the drop does not stick. The condition of small $\Delta\theta$ is difficult to attain, *e.g.* the drop will slide for $\Delta\theta < 1°$, which is very difficult to achieve and almost impossible to maintain. On the contrary, it can be seen that the drop will move if the mean contact angle is greater than 160°. This is due to the reduction of the solid-liquid contact in this limit.

3.3 LIQUID DROPS AS STICKING AGENTS

Menisci form when a small liquid drop is squeezed between two plates as illustrated in Figure 3.3,

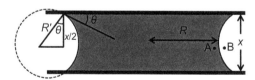

FIGURE 3.3   Illustration of a liquid drop squeezed between two parallel plates separated by a distance x. The drop wets the plates with an equilibrium contact angle θ. The radii of curvature of the menisci is R' while that of the drop after elongation is R. The Laplace pressure between points A and B is ΔP.

giving rise to a capillary bridge, and their curved shapes induce pressure effects and the existence of a Laplace pressure between points A and B induces an interaction between the plates. The two wetted plates can stick together with great strength provided $\theta < 90°$. (This illustration is typical in wet sand where menisci connect different solid sand grains, some of which are platelet.) If the distance x between the plates is very small compared to the radius R of the drop after elongation, the Laplace pressure ΔP can be given as

$$P - P_o = \Delta P = \gamma_{la}\left(\frac{1}{R} - \frac{1}{R'}\right) = \gamma_{la}\left(\frac{1}{R} - \frac{\cos\theta}{x/2}\right) \approx -\frac{2\gamma_{la}\cos\theta}{x} \tag{3.11}$$

in which P_o is the atmospheric pressure, P is the pressure inside the liquid, and R' is the radius of curvature of the menisci. If $x \ll R$, the plates are attracted to each other by a force F,

$$F = \Delta P \times A = \pi R^2 \frac{2\gamma_{la}\cos\theta}{x} \tag{3.12}$$

in which $A = \pi R^2$ is the surface area of the capillary bridge. The force is quite large. For water of $\gamma_{la} = 72$ mN m^{-1} (at 25°C), $R = 1.5$ cm, $x = 6$ μm and $\theta = 0°$, $\Delta P \sim 24$ kPa and $F \sim 17$ N, the weight of ~2 L of water. Such capillary adhesion is used by some insects, *e.g.* the beetle *Hemisphaerota cyanea*, in nature to stick to their substrate. When assaulted, it presses its numerous tarsi on the ground, generating a force close to 9.8×10^{-3} N (~60 times its body weight) for more than 2 min, thereby resisting the attacking ants. A footprint made of many tiny drops is left when the beetle leaves its clinging substrate. This reveals the presence of a liquid (presumably oil) during the contact. The contact angle dictates the type of the interaction. The capillary force is attractive if $\theta < 90°$ and repulsive if $\theta > 90°$.

3.4 NON-STICKING DROPS

On certain natural, *e.g.* the lotus and *Gingko biloba* leaves, and synthetic surfaces, liquid drops roll off (*i.e.* do not stick) when the three-phase contact angle $\theta \gg 90°$ and non-sticking drops are witnessed. In many cases, this happens when θ is between 150° and 180°, although the upper limit is rarely met, and such surfaces are said to be super'phobic, as discussed in Chapter 2. The friction between the liquid drop and surfaces is negligible and drag effects are entirely absent. In addition, the drop has little affinity for the surfaces of the solid.

Leidenfrost and non-coalescing drops and liquid marbles are other examples of non-sticking drops and are discussed in the following sub-sections. Compared to super'phobic surfaces where complete non-wetting (*i.e.* contact angle of 180° and 0° hysteresis) is unachievable, complete non-wetting is achieved with these drops.

3.4.1 LEIDENFROST DROPS

When a liquid drop is deposited on a hot solid surface whose temperature is around the boiling point of the liquid, the drop spreads over the plate in a thin layer, boils, and quickly vanishes. If, however, the temperature of the solid surface is much higher than the boiling point of the liquid, the drop does not make contact with the surface but levitates above its own vapor (thickness typically ~ 100 μm) and remains completely non-wetting. Because of the insulating properties of the film, the drop evaporates slowly. Typically, the lifetime τ of the drop is increased by up to 500 times. For example, a millimetre-sized drop of water on a metallic surface at 200°C is observed to float for more than a minute (Biance *et al.* 2003). Also, the absence of contact between the liquid and the solid prevents the nucleation of bubbles so that the drop does not boil but rather evaporates quietly. These drops (Figure 3.4)

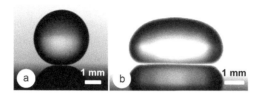

FIGURE 3.4 Photograph of (a) quasi-spherical and (b) puddle Leidenfrost water drops, resting on a cushion of water vapor (thickness ~ 100 μm), on a flat metallic plate maintained at 300°C. (Courtesy of Raphaële Thévenin and Dan Soto.)

are known as *Leidenfrost* drops, named after the German physician, Johann Gottlob Leidenfrost, who first published an extensive report on the phenomenon in 1756 following the first report by Herman Boerhaave in 1732. The temperature at which these drops form is known as the Leidenfrost temperature T. Increasing the temperature of the metal surface decreases the lifetime of the drop because conduction and radiation between the plate and the drop are enhanced. Leidenfrost drops are extremely mobile, and any slope makes them run down, and as a result, they are studied in a ring heated to the same temperature as the plate.

3.4.1.1 Shape and Stability

The temperature inside a Leidenfrost drop is considered to be equal to the boiling temperature of the liquid after a short transient time so that the quantities γ_{la} and ρ are known for a given liquid. Water (boiling point 100°C), $\gamma_{la} = 59$ mN m^{-1} and $\rho = 960$ kg m^{-3}, has $\kappa^{-1} = 2.9$ mm. At equilibrium, the Laplace pressure $\gamma_{la}C$, where C is the curvature of the interface, is balanced by the hydrostatic pressure $\rho g(H - z)$, where z is the vertical coordinate starting from the plate. The equation for the drop shape can be written as $C = C_o + (H - z)/(\kappa^{-1})^2$, with C_o being the top curvature. A Leidenfrost drop is nearly spherical (*i.e.* quasi-spherical), with a flattened bottom or contact where it rests on the solid surface, when its radius R_o is smaller than the capillary length κ^{-1}. The curvature is $C_o \approx 2/R_o$ and height $H \approx 2R_o$. The radius δ of the contact zone is given by a balance between the liquid surface tension and gravity, which gives $\delta \approx R_o^2/\kappa^{-1}$. A puddle-shaped Leidenfrost drop is obtained when the drop radius is larger than κ^{-1}. In this case, the size of the contact zone is approximately equal to the drop radius (*i.e.* $\delta \approx R_o$). The top of a puddle is nearly flat ($C_o \approx 0$) and the shape equation is $C = (H - z)/(\kappa^{-1})^2$. The curvature increases linearly along the interface going from the top ($z = H, C = 0$) to the bottom $\left[z = 0, C = H/(\kappa^{-1})^2\right]$. Upon integrating analytically, the shape equation gives the thickness of a non-wetting gravitational pancake $H = 2\kappa^{-1}$. This result can also be confirmed from the energy E of the puddle, the sum of the gravitational and surface tension components. When $R_o \gg \kappa^{-1}$, the energy can be written as $E \approx \pi\rho g R_o^2 H^2/2 + 2\pi\gamma_{la}R^2 \approx \rho g H\, \Omega/2 + 2\gamma_{la}\Omega/H$, where $\Omega \approx \pi R_o^2 H$ is the volume of the puddle. The minimization of E with respect to H at constant Ω gives $H \approx 2\kappa^{-1}$ just like in the previous case. The transition between spheres and pancakes is observed when the gravitational energy of a sphere is larger than its surface energy. Passing from a small drop to a large one, the liquid H first increases as $2R_o$ (for $R_o < \kappa^{-1}$) before balancing out at $2\kappa^{-1}$ (for $R_o > \kappa^{-1}$) (Biance *et al.* 2003). This result is not valid when the drop size is just above the κ^{-1}. The Laplace pressure associated with the equator curvature tends to squeeze the liquid, resulting in a thickness slightly greater than $2\kappa^{-1}$. It has been shown (Aussillous and Quéré 2006) numerically that the maximum thickness, $H = 2.1\kappa^{-1}$, is obtained when $R_o = 3.2\kappa^{-1}$.

The effects of gravity can be neglected if the drop radius is smaller than the capillary length, but small drops are also flattened by gravity. This usually happens close to the substrate. A small deformation ς of a sphere of radius R_o contacting a non-deformable substrate induces a contact zone of radius δ. From elementary geometry, $\delta \sim (R_o\varsigma)^{0.5}$ and increases at small ς. The deformation results from the drop weight, $\sim \rho g R_o^3$, and it is limited by an elastic force $\gamma_{la}\varsigma$. Thus, a vertical deformation is approximately $R_o^3/(\kappa^{-1})^2$ and a lateral contact scales as R_o^2/κ^{-1}. The quadratic law of contact $\sim R_o^2/\kappa^{-1}$ is very different from what is true for a wetting drop, which increases linearly with the drop size $\sim R\sin\theta$. This particularly implies that the smaller the drop, the larger the Laplace pressure exerted on the subjacent film. Because of this divergence, a Leidenfrost drop will only contact its hot substrate at the moment it "vanishes".

Although static arguments are used above, both the liquid and vapor move. As convection takes place inside the drop (viscosity η), the vapor film (viscosity η_v), pressed by the liquid drop, escapes laterally. These flows affect the shape of the liquid drop, although quite marginally. The flow in the drop is caused by different factors. For example, as liquid is drawn by the moving vapor, the viscous stress $\eta_v U/\tau$ in the film balances the viscous stress $\eta V/R_o$ in the drop giving rise to a typical velocity $V \sim U(R_o/\tau)(\eta_v/\eta)$. The $V \sim 1$ cm s^{-1} for a vapor velocity $U = 10$ cm s^{-1} and film thickness $\tau = 100$ μm. The temperature in the liquid also decreases by a few degrees from the film, where it is at the boiling point, generating Marangoni flow. Balancing the viscous stress $\eta V/R_o$ with the gradient of surface tension $\Delta\gamma_{la}/R_o$ along the drop gives $V \sim \Delta\gamma_{la}/\eta$, which is a typical Marangoni velocity in the liquid. For water of $\Delta\gamma_{la} \approx 10^{-4}$ mN m^{-1} and a temperature difference of a few degrees, $V \sim 10$ cm s^{-1}, typical of the liquid velocity inside a Leidenfrost drop. The corresponding Weber and capillary numbers are small compared to 1, meaning that the static shape is still maintained by the surface tension despite the presence of these flows.

3.4.1.2 The Vapor Layer: Stationary States

Leidenfrost drops levitate and press on the vapor they produce and this tunes the thickness τ of the vapor film. The vapor layer or film extends the lifetime of a Leidenfrost drop. Its thickness is measured accurately by diffraction of He-Ne laser beam by the slit between the drop and the surfaces of the solid substrate. The film thickness is computed from the diffraction pattern, and values in the range of 10–100 μm have been reported (Biance *et al.* 2003, Myers and Charpin 2009). Because the film thickness may vary with time during its measurement, liquid is normally supplied from the outside into the drop under investigation and the drop radius is determined by fixing the feeding rate. In the stationary state or regime, the vapor film is created by drop evaporation, but the drop presses on it and causes it to flow laterally. The film flow rates caused by evaporation and the pressure supplied by the drop's weight can be evaluated. Consider a Leidenfrost drop ($R_0 \sim \kappa^{-1}$) for which the bottom is fairly flat, and its shape is comparable to that of a disk of radius R'' and thickness κ^{-1}. The exchange of heat energy between the solid surfaces and the liquid drop, across the vapor gap, is mainly through conduction. The corresponding heat flux per unit area is $\lambda(T_s - T_b)/\tau = \lambda \Delta T/\tau$. The $\Delta T = T_s - T_b$ is the temperature difference between the temperature T_s of the solid and boiling point T_b of the liquid and λ is the thermal conductivity of the vapor. [The Stefan's law, $\sigma(T_s^4 - T_b^4)$, in which σ is the Stefan's constant, gives the radiation heat flux per unit area. The ratio of the radiative flux to the convective flux at $\sim 300°C$ is around 0.05. Both fluxes are approximately equal above 1000°C, indicative that the heat flux at lower solid temperatures is dominated by conduction.] Therefore, the flow or evaporation rate $\dot{m}$, *i.e.* the mass of liquid evaporated per unit time, can be deduced. First, the quantity Q of heat energy supplied to the liquid per unit time is proportional to the surface area $\pi\delta^2$ of the contact zone, the λ, and the temperature gradient $\Delta T/\tau$. In the stationary state, where the liquid temperature equals its boiling temperature, all the heat supplied is used for evaporation and the evaporation rate is

$$\dot{m} = \frac{dm}{dt} = \frac{\lambda \Delta T}{L\tau}\pi\delta^2 \tag{3.13}$$

The L is the latent heat of evaporation, m is mass of the liquid drop and t is the time. Second, the liquid drop presses on the vapor film and causes it to escape laterally in Poiseuille flow (assuming a non-slip condition on both the solid and liquid interfaces). Because of the vapor film geometry (few millimetres long and small thickness ~ 0.1 mm), the lubrication approximation can be used. The relationship between the vapor flux and the pressure gradient ΔP responsible for the flow can be obtained from the flow rate given as

$$\dot{m} = \frac{dm}{dt} = \frac{2\pi\rho_v\tau^3}{3\eta_v}\Delta P \tag{3.14}$$

where ρ_v is the vapor density.

Recalling that for small drops, $R_0 < \kappa^{-1}$ and $\delta \approx R_0^2/\kappa^{-1}$, the pressure ΔP acting on them or pressure gradient is the Laplace pressure $2\gamma_{la}/R_0$ gives

$$\dot{m} = \frac{dm}{dt} = \frac{2\pi\rho_v\tau^3}{3\eta_v} \times \frac{2\gamma_{la}}{R_0} \tag{3.15}$$

However, in the stationary state, the film is replenished by evaporation at the same rate the vapor escapes and a law for the film thickness can be obtained by equating Equations (3.13) and (3.15). This gives

$$\tau = \left(\frac{3}{4}\frac{\lambda\Delta T\eta_v}{L\rho_v\gamma_{la}(\kappa^{-1})^2}\right)^{\frac{1}{4}} \times R_o^{5/4} \Rightarrow \tau \sim \left(\frac{\lambda\Delta T\eta_v}{L\rho_v\gamma_{la}(\kappa^{-1})^2}\right)^{\frac{1}{4}} \times R_o^{5/4} \tag{3.16}$$

Equation (3.16) shows that τ varies as $R_o^{5/4}$. However, for very small drops, the film is expected to play a minor role in the evaporation process because its surface area $(\pi\delta^2)$ vanishes rapidly as R_o^4. The temperature gradient is expected to be $\Delta T/R_o$ and evaporation takes place over the entire drop surface and the rate of evaporation is

$$\dot{m} \sim \frac{\lambda}{L}\frac{\Delta T}{R_o}\pi\frac{R_o^4}{(\kappa^{-1})^2} \tag{3.17}$$

The rate is larger than the one given in Equation (3.13) if R_o is smaller than $[(\kappa^{-1})^2\tau]^{1/3}$, e.g. millimetric drops or less. The proportion of vapor which replenishes the vapor layer scales as the surface area ratio τ^2/R_o^2. Equating Equations (3.15) and (3.16) gives the film thickness as

$$\tau = \left(\frac{3}{4}\frac{\lambda\Delta T\eta_v\rho g}{L\rho_v\gamma_{la}^2}\right)^{\frac{1}{3}}R_o^{4/3} \Rightarrow \tau \sim \left(\frac{\lambda\Delta T\eta_v\rho g}{L\rho_v\gamma_{la}^2}\right)^{\frac{1}{3}}R_o^{4/3} \tag{3.18}$$

For puddle Leidenfrost drops, however, $R_o > \kappa^{-1}$ and $\delta \sim R_o$, and the pressure applied on the film is $2\rho g\kappa^{-1}$, and Equation (3.14) becomes

$$\dot{m} = \frac{dm}{dt} = \frac{2\pi\rho_v\tau^3}{3\eta_v}\times 2\rho g\kappa^{-1}, \text{ where } 2\rho g\kappa^{-1} = \Delta P \tag{3.19}$$

The film thickness, as obtained by equating Equations (3.13) and (3.19), is

$$\tau = \left(\frac{3}{4}\frac{\lambda\Delta T\eta_v}{L\rho_v\rho g\kappa^{-1}}\right)^{\frac{1}{4}}R_o^{1/2} \Rightarrow \tau \sim \left(\frac{\lambda\Delta T\eta_v}{L\rho_v\rho g\kappa^{-1}}\right)^{\frac{1}{4}}R_o^{1/2} \tag{3.20}$$

Equations (3.16 through 3.20) show that the film thickness, generally, increases monotonically with the drop radius but the increment depends on the drop size and many experiments have confirmed this. The combination of Equations (3.16 through 3.20) with Equation (3.13) permits the evaluation of the quantity of vapor produced per unit time. This also provides a measure of the lifetime $\iota \sim m/\dot{m}$ of the drop, provided it evaporates mainly through its bottom. The mean vapor velocity U is obtained from the expression

$$\dot{m} \approx 2\pi R_o\rho_v U \tag{3.21}$$

The Reynolds number R_e, obtained by comparing inertial and viscous effects in the film, is found to be

$$R_e \approx \frac{\rho_v \tau^2 U}{R_o \eta_v} \tag{3.22}$$

Example 3.3: Estimating Film Thickness

Estimate the film thickness for a puddle water drop (radius 5 mm) placed on a hot metallic plate maintained at 300°C given that the density of water at its boiling point (100°C) is ~ 958 kg m^{-3} and the surface tension is 59 mN m^{-1} and the latent heat of evaporation is 2.26×10^6 J kg^{-1}, while the thermal conductivity, density, and viscosity of water vapor (100°C) are 0.025 W m^{-1}K^{-1}, 1 kg m^{-3}, and 2×10^{-5} Pa s, respectively.

Method

Substitute the given quantities into Equation (3.20), bearing in mind that the κ^{-1} at 100°C is ~ 2.5 mm, and solve for the film thickness at once.

Answer

$$\tau \sim \left(\frac{\lambda \Delta T \eta_v}{L \rho_v \rho g \kappa^{-1}} \right)^{\frac{1}{4}} R_o^{1/2}$$

$$\sim \left(\frac{0.025 \text{ W m}^{-1}\text{ K}^{-1} \times 473.15 \text{ K} \times 2 \times 10^{-5} \overbrace{\text{kg m}^{-1}\text{ s}^{-2}}^{\text{Pa}} \text{ s}}{2.26 \times 10^6 \text{ J kg}^{-1} \times 1 \text{ kg m}^{-3} \times 958 \text{ kg m}^{-3} \times 9.8 \text{ m s}^{-2} \times 2.5 \times 10^{-3} \text{ m}} \right)^{\frac{1}{4}} \times \sqrt{5 \times 10^{-3} \text{ m}}$$

$$\sim 102.76 \times 10^{-6} \text{ m} \sim 103 \text{ μm}$$

3.4.1.3 Evaporation of Leidenfrost Drops

Evaporation of a Leidenfrost drop is followed by noting its radius as a function of time. The evaporation depends on the size of the drop, e.g. for puddles, $\tau \propto R_o^{1/2}$ and $\dot{m} \propto R_o^{3/2}$, and the drops sink in the vapor as evaporation proceeds. During evaporation, the drop radius decreases regularly, except at the end where the drop becomes quasi-spherical and the variation becomes quicker. The evaporation also increases with increasing temperature of the solid surfaces, leading to a smaller drop lifetime. The film thickness also decreases with evaporation. Generally, both the contact zone and the film thickness decreases and vanishes as the drop evaporates. The drop disappears when τ and R_o cancel out. Consider the evaporation of a puddle Leidenfrost drop. The radius of the evaporating puddle and the thickness of the vapor film are related through Equation (3.20). The time dependence of the radius can be obtained by substituting Equation (3.20) into (3.19), provided the evaporation is dominated by the vapor film. If R_o' is the drop radius at $t = 0$, the radius $R_o(t)$ at time t would be

$$R_o(t) = R_o\left(1-\frac{t}{\iota}\right)^2 \quad \iota = 2\left(\frac{4\rho\kappa^{-1}L}{\lambda\Delta T}\right)^{3/4}\left(\frac{3\eta_v}{\rho_v g}\right)^{1/4}(R_o')^{1/2} \tag{3.23}$$

This equation agrees very well with experimental data, provided the drop is a puddle, and is used for predicting the lifetime of the drops which is found to decrease as $\Delta T^{-3/4}$. The rate at which the thickness of the vapor film evolves with time, Equation (3.24), is obtained by substituting Equation (3.23) into (3.20).

$$\tau(t) = \left(\frac{3}{4}\frac{\lambda\Delta T\eta_v(R_o')^2}{L\rho_v\rho g\kappa^{-1}}\right)^{\frac{1}{4}}\left(1-\frac{t}{\iota}\right) \tag{3.24}$$

Therefore, the thickness of the vapor film is expected to vary linearly with time, vanishing as the drop collapses (for $t = \iota$). This agrees very well with experimental data (Biance *et al.* 2003).

For smaller drops, where evaporation occurs through the whole drop surface, the analogue of Equation (3.23) is

$$R_o(t) = R_o\left(1-\frac{t}{\iota}\right)^{\frac{1}{2}} \quad \iota = \frac{\rho L}{\lambda\Delta T}(R_o')^2 \tag{3.25}$$

Equation (3.25) implies that the rate of drop retraction increases close to its vanishing time, which is in qualitative agreement with experimental observations. In addition, this equation provides a measure of the lifetime of small Leidenfrost drops, which is found to be slightly more sensitive to temperature compared with a puddle.

3.4.1.4   Self-propelled Leidenfrost Drops: Self-propelling Force and Friction

In addition to imparting long lifetime and an almost frictionless motion to Leidenfrost drops, the vapor layer, through its flow, also propels the drops when placed on a hot textured solid surface (Lagubeau *et al.* 2011). Levitating Leidenfrost drops self-propel on textured solid surfaces with asymmetric teeth, provided the teeth are at a much higher temperature than the boiling point of the liquid (Linke *et al.* 2006). Consider the self-propulsion of a flattened puddle Leidenfrost drop on a textured solid surface as shown in Figure 3.5 (upper image). Let the drop be in the stationary state where its temperature is the boiling temperature of the liquid. The vapor escapes isotropically, but the texture breaks the symmetry of the surface and directs the vapor flow which can be asymmetric. The vapor flow is considered irreversible because the Reynolds number for the flow is of the order of 10. As the vapor moves towards the textured surface, the flow resistance is higher than in the reverse direction. As a result, the vapor escapes along the smallest slopes of the texture and this propels the drop in the forward direction. In other words, the texture converts a uniform vapor flow into a jet thrust.

Equation (3.19) gives the mass $\dot{m}$ ejected per unit time for a puddle Leidenfrost drop and the force propelling the drop scales as $\dot{m} \times \Delta u$ where Δu is the velocity difference of the vapor flow between two opposite directions. The Δu increases with

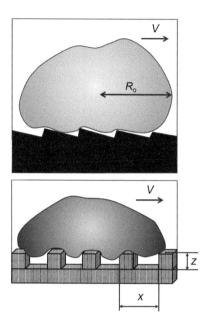

FIGURE 3.5 The upper image shows the schematic of a self-propelling puddle Leidenfrost drop of radius R_o moving, in the right direction with velocity V, on a rough solid surface. The lower image shows the schematic of a puddle Leidenfrost drop moving on a rough solid surface (at high temperature) where the length and depth of each tooth is x and z, respectively. The textured surface distorts the drop where it contacts the surface.

drop asymmetricity and decreases with the flow thickness τ. The Δu is also expected to increase proportionally with the velocity of the drop. For puddle drops, the radius is approximately equal to the radius of the contact zone and according to the law of conservation of matter,

$$\dot{m} = 2\pi\tau\rho_v u R_o \tag{3.26}$$

Also, Equation (3.19) combines with (3.20) to give $\dot{m}$ as

$$\dot{m} = \rho_v \frac{\alpha^{\frac{3}{2}} R_o^{\frac{3}{2}}}{\beta} \text{ where } \alpha = \left(\frac{3\lambda\Delta T\eta_v}{4L\rho_v\rho g\kappa^{-1}} \right)^{\frac{1}{2}} \text{ and } \beta = \frac{3\eta_v}{2\pi\rho g H} \tag{3.27}$$

The combination of Equations (3.26) and (3.27) gives u as

$$u = \frac{\alpha^{\frac{3}{2}} R_o^{\frac{1}{2}}}{2\pi\tau\beta} \tag{3.28}$$

Substituting u into Equation (3.26) gives the self-propelling force $F_p \approx \dot{m}u$ as

$$F_p \approx F_0 \times \left(\frac{R_0}{\alpha}\right)^{\frac{3}{2}} \text{ where } F_0 = \frac{\rho\alpha^4}{\beta^2} \tag{3.29}$$

The self-propelling force is of the order of 10 µN, and it increases rapidly with the drop radius in line with Equation (3.29). The force is small compared with other static forces, but it is large enough to induce the fast motion of the drop because of the low fraction provided by the vapor layer.

Example 3.4: Estimating the Self-propelling Force

Estimate the self-propelling force for the drop described in Example 3.3.

Method

Substitute the given quantities into Equation (3.29) and solve for the self-propelling force F_p at once.

Answer

$$F_p \approx F_0 \times \left(\frac{R_0}{\alpha}\right)^{\frac{3}{2}} \text{ where } F_0 = \frac{\rho\alpha^4}{\beta^2}, \alpha = \left(\frac{3\lambda\Delta T\eta_v}{4L\rho_v\rho g\kappa^{-1}}\right)^{\frac{1}{2}} \text{ and } \beta = \frac{3\eta_v}{2\pi\rho gH}$$

$$\alpha = \left(\frac{3 \times 0.025 \overbrace{\text{J s}^{-1} \text{ m}^{-1} \text{ K}^{-1}}^{W} \times 473.15 \text{ K} \times 2 \times 10^{-5} \overbrace{\text{kg m}^{-1} \text{ s}^{-2}}^{Pa} \text{ s}}{4 \times 2.26 \times 10^6 \text{ J kg}^{-1} \times 1 \text{ kg m}^{-3} \times 958 \text{ kg m}^{-3} \times 9.8 \text{ m s}^{-2} \times 2.5 \times 10^{-3} \text{ m}}\right)^{\frac{1}{2}}$$

$$= 1.83 \times 10^{-6} \text{ m}$$

$$\beta = \frac{3 \times 2 \times 10^{-5} \overbrace{\text{kg m}^{-1} \text{ s}^{-2}}^{Pa} \text{ s}}{2\pi \times 958 \text{ kg m}^{-3} \times 9.8 \text{ m s}^{-2} \times 2 \times 2.5 \times 10^{-3} \text{ m}} = 2.03 \times 10^{-7} \text{ s}$$

$$F_0 = \frac{\rho\alpha^4}{\beta^2} = \frac{958 \text{ kg m}^{-3} \times (1.83 \times 10^{-6})^4 \text{ m}^4}{(2.03 \times 10^{-7})^2 \text{ s}^2} = 2.6 \times 10^{-7} \text{ N}$$

$$F_p \approx 2.6 \times 10^{-7} \text{ N} \times \left(\frac{5 \times 10^{-3} \text{ m}}{1.83 \times 10^{-6} \text{ m}}\right)^{\frac{3}{2}} \approx 37.13 \times 10^{-3} \text{ N} \approx 37.13 \text{ mN}$$

The viscous friction in a thin vapor film scales as $(\eta_v v/\tau)R_0^2$, in which v is the drop velocity, while the air-drag force scales as $\rho v^2 R_0^2$. The magnitude of both forces is of the order of 0.1 µN, which is 100 times smaller than the driving force F_p. In the stationary state, the propulsion and frictional forces balance each other out and the drop moves with a terminal velocity V. This implies that the drop experiences an additional friction whose magnitude is higher than the previous one. This is due to the surface texture (*i.e.* the teeth on the solid surfaces). Rolls of drops hit the teeth as the drop moves and energy is dissipated and the corresponding frictional force F'

can be evaluated. Let x and z be the length and depth of each tooth respectively as illustrated in Figure 3.5 (lower image). The mass per roll scales as $\rho\, x\, zR_0$ while the loss of kinetic energy per roll scales as $\rho\, x\, zR_0 v^2$. If R_0/x is the number of bumps, the total loss in kinetic energy $\rho z R_0^2 v^2$ is the product of the loss of kinetic energy per roll and the number of bumps and represents the work $F'x$ of a frictional force while the frictional force itself is given by

$$F' \approx \rho v^2 R_0^2 \frac{x}{z} \qquad (3.30)$$

Where the terminal velocity v of the drop can be calculated using Equation (3.31).

$$v \approx \frac{\alpha}{\beta}\left(\frac{\rho_v}{\rho}\right)^{0.5}\left(\frac{x}{z}\right)^{0.5}\left(\frac{\alpha}{R_0}\right)^{0.25} \qquad (3.31)$$

Equation (3.31) shows that v is almost insensitive to the drop radius but depends strongly on the geometry of the rough solid surface. Equation (3.30) shows that friction would be enhanced using a rough surface with a larger x and a larger z. Such surfaces turn to decelerate the drop compared to flat smooth ones and can be used for trapping fast moving Leidenfrost drops. Experimentally, F' is observed to be of the order of 10 μN. If the liquid drop is replaced by a solid (e.g. dry ice) that does not deform on the textured solid surface, friction will be reduced while the terminal velocity will increase.

3.4.1.5 Deceleration and Trapping of Leidenfrost Drops

Rough surfaces with asymmetrical teeth are known to enhance the self-propelling force, and the drops naturally accelerate. Surfaces with symmetric teeth turn to decelerate and trap Leidenfrost drops, as confirmed by Dupeux et al. (2011). Using hot (temperature 450°C) rough solid surfaces with $x = 3$ mm and $z = 480$ μm, Figure 3.5 (lower image), it was observed (Dupeux et al. 2011) that drop deceleration was faster compared with a flat smooth solid surface. Two deceleration regimes were witnessed. The drop velocity decreases exponentially with the distances reaching a certain value after which it drastically decreases to zero. This shows that textured surfaces have the capacity to trap Leidenfrost drops. The frictional force, however, was found to lie between 20 and 300 μN.

3.4.2 Non-coalescing Drops

Two drops of the same liquid or miscible liquids or a drop and its bulk liquid (or miscible liquid) phase are expected to coalesce (i.e. "unite") once in apparent contact with each other. However, it has been observed that this does not occur in many cases. Usually, a liquid drop placed on the surface of its bulk liquid phase takes as much as a fraction of a second to coalesce. Similarly, two drops of the same liquid take a similar time frame to coalesce once in apparent contact with each other. For example, splashing water in a container or on a river when paddling a canoe produces drops that float on the surface for several seconds before uniting with the bulk liquid phase. Also, hot coffee dripping from a drip-style coffeemaker generally produces drops that spend several seconds on the surface of the bulk coffee before sinking in it. In another instance, a drop that normally wets a solid surface may deform against it rather than wet it. These drops are said to be non-coalescing and were first reported by Reynolds (1881). Drop non-coalescence may be temporary or permanent and has been adduced to the existence of a layer of intervening substances (a liquid or a gas) between the two phases. It has also been observed that the temperature differences between a liquid drop and a bulk liquid phase also induces non-coalescence due to thermal Marangoni motion, owing to the surface tension gradient at the pool surface (Savino *et al.* 2003). When the liquid drop (colder) makes contact with the bulk liquid surface (hotter), there is radial flow towards the drop. The radial flow field drags the ambient air under the drop, creating an air film which prevents a direct contact between the drop and the bulk liquid molecules (Savino *et al.* 2003). This is akin to what happens when a surfactant-laden water drop makes an apparent contact with an aqueous surfactant solution where the friction between air and the surfactant layer inhibits the thinning of the air film, thereby preventing coalescence (Amarouchene *et al.* 2001) as shown in Figure 3.6.

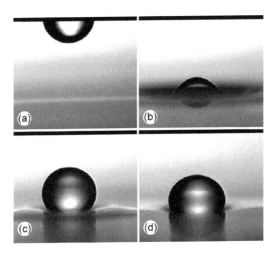

FIGURE 3.6 Photograph of a surfactant drop: (a) falling, (b) impacting a bulk surfactant solution, (c) rebounding after impact, and (d) resting on the solution surface before coalescence. (From Amarouchene, Y. *et al.*, *Phys. Rev. Lett.*, 87, 206104, 2001.)

3.4.3 LIQUID MARBLES

We have seen in Chapter 2 that when a small liquid drop is deposited on a solid surface, it either forms a spherical cap shape with a well-defined equilibrium contact angle to the solid or spreads across the surface until it forms a wetting film (de Gennes 1985, Leger and Joanny 1992, McHale *et al.* 2004). Depending on the magnitude of the equilibrium advancing contact angle, solid particles and surfaces were said to be "philic" or "phobic". Here and in Chapters 4 and 5, the focus is on phobic powdered particles where the equilibrium particle-liquid-air three-phase contact angle is above 90°. These particles possess relatively low surface energy.

It has been shown (Aussillous and Quéré 2001, 2006, Quéré and Aussillous 2002), experimentally, that hydrophobic powdered particles can spread over water creating a surface film, in contrast to the long-held knowledge that liquids spread on solids (but not the other way round). Their pioneering work has intrigued and formed the basis of liquid marble research. Liquid marbles (Figure 3.7) are nonstick millimetre-sized liquid-in-air drops, encapsulated by microparticles or nanoparticles, which demonstrate very low friction when rolling on solid substrates (Aussillous and Quéré 2001, 2006, Quéré and Aussillous 2002, Vella *et al.* 2004). The particles, which are either hydrophobic, oleophobic, superhydrophobic, superoleophobic, or omniphobic depending on the liquid, poorly wet the liquid drop and protrude largely into the air phase. Liquid marbles are separated from a solid or liquid support by air pockets, similar to Leidenfrost drops, and this accounts for their inherent non-Amontonian friction.

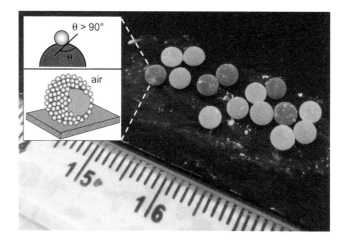

FIGURE 3.7 Photograph of liquid marbles of α-hexylcinnamaldhyde (orange) and squal-ane (red and white) stabilized by fluorinated PFX 10-ZnO particles. Some of the squal-ane liquid marbles are red because Sudan red dye (0.4 mg) was added to the precursor liquid (5 cm^3) before formation of the marbles. Photograph courtesy of the author. Inset: Schematic of a liquid marble, showing particles on its surfaces and the possible equilibrium contact angle θ.

Typical examples of liquid marbles abound in nature. The waste disposal protocol of galling aphids ends up in the formation of liquid marbles. Phloem-feeders excrete large volumes of honeydew waste in which young or immobile insects can become stuck or asphyxiated. As the honeydew is defecated, it is covered with a powdery wax secreted from the anus by the aphid, thus creating a wax-covered sphere which the soldier aphids then roll out of a hole (Pike *et al.* 2002). The principle being used here is a simple one of encapsulating a liquid entirely in a hydrophobic powder. Liquid marble formation also occurs after rain as "summer ice" (Aussillous and Quéré 2001), after wildfire creates hydrophobic soil (McHale *et al.* 2007a, 2007b), and in the wet granulation of a highly hydrophobic fine powder (Hapgood and Khanmohammadi 2009, McEleney *et al.* 2009).

3.5   CONCLUSION

The quasi-spherical and puddle shapes of sessile liquid drops are discussed in terms of the interplay between gravity and liquid surface tension. The discussion is extended to sticking drops, with rain drops on vertical glass windows or panes, car windscreens and other inclined surfaces as modeled examples. Next, liquid drops were viewed as sticking agents as they are able to hold two parallel flat plates, separated by a small distance, containing them. Lastly, non-sticking drops, with Leidenfrost drops as typical examples, non-coalescence drops, with liquid marbles as typical examples, were discussed with the aim of setting the scene for Chapter 4.

EXERCISES

Discussion Questions

Question 1

a. Why are some sessile liquid drops quasi-spherical and some puddle-shaped?
b. Describe how the beetle (*Hemisphaerota cyanea*) opposes attack by other insects.
c. Differentiate between sticking and non-sticking drops.

Question 2

a. What are Leidenfrost drops?
b. How are Leidenfrost drops able to self-propel on textured solid surfaces?

Question 3

a. What are non-coalescence drops?
b. What are liquid marbles?
c. Differentiate between Leidenfrost drops and liquid marbles.

Numerical Questions

Question 1

a. Show that the capillary length κ^{-1} of a liquid can be written as $(\gamma_{la}/\rho g)^{0.5}$. Calculate the capillary length of glycerol, whose surface tension and density at 20°C are 63.4 mN m^{-1} and 1.26 g cm^{-3}, respectively.
b. Show that the force of attraction F between two flat plates separated by a distance x, forming a capillary bridge of radius R and equilibrium contact angle θ, is $\pi R^2 2\gamma_{la} \cos\theta / x$.
c. What magnitude of force will push two parallel glass slides, with a water drop ($\gamma_{la} = 75$ mN m^{-1}, radius 0.2 cm) in between, separated by 1 μm if the contact angle between the water and the glass plate is 65°?

Question 2

a. Starting with

$$E = (\gamma_{sl} - \gamma_{sa})A + \gamma_{la}A + \frac{1}{2}\rho g H^2 A$$

the total energy of a puddle Leidenfrost drop, in which H is the puddle height, ρ is liquid density, and g is the acceleration of gravity, A represents the area of contact between the puddle and the substrate, γ_{sl}, γ_{sa} and γ_{la} represent the solid-liquid, solid-air, and liquid-air interfacial tensions, respectively. Show that

$$E \approx \pi \rho g R_o^2 H^2 / 2 + 2\pi\gamma_{la}R^2 \approx \rho g H\,\Omega/2 + 2\gamma_{la}\Omega/H$$

where $\Omega \approx \pi R_o^2 H$ is the puddle volume.

b. Show that the minimization of $E \approx \rho g H\, \Omega/2 + 2\gamma_{la}\Omega/H$ with respect to H at constant Ω gives $H \approx 2\kappa^{-1}$.

Question 3

a. Show that for small and puddle Leidenfrost drops, the film thickness is a monotonic function of the drop radius.
b. Show that the self-propelling of Leidenfrost drop on textured solid surfaces is proportional to the cube root of the drop radius.
c. For a puddle Leidenfrost drop (radius 4 mm) existing on a partially rough hot metallic plate maintained at 350°C, estimate the film thickness and the self-propelling force given that the density of water at its boiling point (100°C) is ~ 958 kg m^{-3} and the surface tension is 59 mN m^{-1} and the latent heat of evaporation is 2.26×10^6 J kg^{-1} while the thermal conductivity, density, and viscosity of water vapor (100°C) are 0.025 W m^{-1} K^{-1}, 1 kg m^{-3} and 2×10^{-5} Pa s, respectively.

FURTHER READING

Quéré, D. "Non-sticking Drops." *Rep. Prog. Phys.* 68 (2005): 2495–532.
Quéré, D. "Leidenfrost Dynamics." *Annu. Rev. Fluid Mech.* 45 (2013): 197–215.
Sobac, B., A. Rednikov and S. Dorbolo. "Leidenfrost Drops". In *Droplet Wetting and Evaporation.* San Diego: Elsevier, 2003.

REFERENCES

Amarouchene, Y., G. Cristobal and H. Kellay. "Noncoalescence." *Phys. Rev. Lett.* 87 (20) (2001): 206104.
Aussillous, P. and D. Quéré. "Liquid Marbles." *Nature* 411 (2001): 924–27.
Aussillous, P. and D. Quéré. "Properties of Liquid Marbles." *Proc. R. Soc. Chem. Lond. A* 462 (2006): 973–99.
Biance, A.L., C. Clanet and D. Quéré. "Leidenfrost Drops." *Phys. Fluids* 15 (2003): 1632.
de Gennes, P.G. "Wetting: Statics and Dynamics." *Rev. Mod. Phys.* 57 (3) (1985): 827–63.
Dupeux, G., M.L. Merrer, C. Clanet and D. Quéré. "Trapping Leidenfrost Drops with Crenelations." *Phys. Rev. Lett.* 107 (2011): 114503.
Hapgood, K.P. and B. Khanmohammadi. "Granulation of Hydrophobic Powders." *Powder Technol.* 189 (2009): 253–62.
Lagubeau, G., M.L. Merrer, C. Clanet and D. Quéré. "Leidenfrost Drops on a Ratchet." *Nat. Phys.* 7 (2011): 1925.
Leger, L. and J.F. Joanny. "Liquid Spreading." *Rep. Prog. Phys.* 55 (4) (1992): 431.
Linke, H., B.J. Aleman, L.D. Melling *et al.* "Self-Propelled Leidenfrost Droplets." *Phys. Rev. Lett.* 96 (2006): 154502.
McEleney, P., G.M. Walker, I.A. Larmour and S.E.J. Bell. "Liquid Marble Formation Using Hydrophobic Powders." *Chem. Eng. J.* 147 (2–3) (2009): 373–82.
McHale, G., N.J. Shirtcliffe and M.I. Newton. "Contact-Angle Hysteresis on Super-Hydrophobic Surfaces." *Langmuir* 20 (23) (2004): 10146–49.
McHale, G., N.J. Shirtcliffe, M.I. Newton and F.B. Pyatt. "Implications of Ideals on Superhydrophobicity for Water Repellent Soils." *Hydrol. Processes* 21 (2007a): 2229–38.
McHale, G., N.J. Shirtcliffe, M.I. Newton, F.B. Pyatt and S.H. Doerr. "Self-Organization of Hydrophobic Soil and Granular Surfaces." *Appl. Phys. Lett* 90 (2007b): 054110.
Myers, T.G. and J.P.F. Charpin. "A Mathematical Model of the Leidenfrost Effect on an Axisymmetric Droplet." *Phys. Fluids* 21 (2009): 063101.
Pike, N., D. Richard, R. Foster and L. Mahadevan. "How Aphids Lose Their Marbles." *Proc. R. Soc. Lond. B* 269 (2002): 1211–15.
Quéré, D. and P. Aussillous. "Non-Stick Droplets." *Chem. Eng. Technol.* 25 (2002): 925–28.
Reynolds, O. "On Drops Floating on the Surface of Water." *Chem. News* 44 (1881): 211.
Savino, R., D. Paterna and M. Lappa. "Marangoni Flotation of Liquid Droplets." *J. Fluid Mech.* 479 (2003): 307–26.
Taylor, G.I. and D.H. Michael. "On Making Holes in a Sheet of Fluid." *J. Fluid Mech.* 58 (4) (1973): 625–39.
Vella, D., P. Aussillous and L. Mahadevan. "Elasticity of an Interfacial Particle Raft." *Europhys. Lett.* 68 (2004): 212–18.

4 Principles and Properties of Liquid Marbles

4.1 PRINCIPLES OF LIQUID MARBLES

Powdered particles (nm to μm) of suitable wettability (three-phase contact angle $\theta \gg 90°$) self-attach and self-assemble on liquid drop surfaces when the liquid drop is rolled on a powdered bed of the particles, encapsulating the liquid drop and leading to the formation of a liquid marble and minimization of the surface energy of the particles, as illustrated in Figures 4.1a,b.

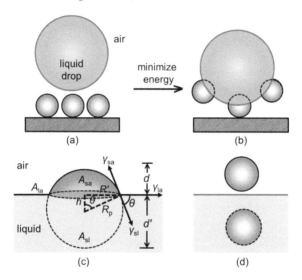

FIGURE 4.1 Illustration of the self-adsorption of powdered particles on the surfaces of a liquid drop leading to minimization of particle surface energy: (a) liquid drop approaching a powdered bed of particles, (b) particles adsorb on the drop surfaces upon arriving the particle bed, and (c) a powdered particle at a liquid-air interface upon adsorption. The particle can be detached and immersed fully into the air or liquid phase as shown in (d).

The particles form either a loosely or closely packed monolayer or multilayer around the liquid drop, with a large fraction protruding out of the liquid drop into the air phase. The packing of the particle layer depends on whether the particle interactions are repulsive or attractive. Attractive particle interactions favor close-packing, while repulsive ones do not, and loosely packed particle layers are formed. For smooth and spherical solid particles of radius R_p forming a monolayer on the liquid drop surfaces, the particles protrude into the air and liquid phases by a distance d and d' (Figure 4.1c), respectively, given as

$$d = R_p - h = R_p(1 - \cos\theta) \quad \text{and} \quad d' = R_p + h = R_p(1 + \cos\theta) \tag{4.1}$$

in which θ is the three-phase contact angle and h is the height of particle centre of mass below the interface. The radius R' of the three-phase contact line is $R_p \sin\theta$. Suffice it to say that the greater the contact angle, the larger the particles protrude out of the drop into the air phase and the larger the separation of the encapsulated liquid from any substrate.

Example 4.1: Calculating the Length of Particle Protrusion into the Air and Liquid Phases for a Liquid Marble

The three-phase contact angle of a powdered particle (diameter 10 μm), adsorbing from air onto the surfaces of a water drop (40 μL) is about 130°. Calculate the length of protrusion of the particle into the air and liquid drop phases.

Method

Substitute $R_p = 5$ μm and $\theta = 130°$ into Equation (4.1) and calculate the distance of particle protrusion into the phases.

Answer

$$d = R_p(1 - \cos\theta) = 5 \text{ μm } (1 - \cos 130°) = 8.2 \text{ μm}$$

$$d' = R_p(1 + \cos\theta) = 5 \text{ μm } (1 + \cos 130°) = 1.8 \text{ μm}$$

During particle attachment, a spherical cap-shaped portion of the liquid drop, of liquid-air interfacial area A_{la}, and the particle's solid-air interfacial area A_{sa} are replaced by the solid-liquid interfacial area A_{sl}. For powdered particles trapped at an interface for which the Bond number Bo, ratio of gravity forces to surface tension forces, Equation (4.2), where $\Delta\rho$ is density of particle minus density of liquid, is far-far less than 1, the surface tension forces dominate.

$$\text{Bo} = \frac{\Delta\rho g R_p^2}{\gamma_{la}} \tag{4.2}$$

However, the gravity forces dominate when Bo is more than 1. When the surface tension forces dominate, the interface is approximately planar in the vicinity of the particle (Figure 4.1c). The surface free energy G_1 for such a particle at the interface (liquid-air in this case) is given by Equation (4.3).

$$G_1 = \gamma_{la}(A_{la} - \pi R'^2) + \gamma_{sl} A_{sl} + \gamma_{sa} A_{sa} \tag{4.3}$$

The γ_{la}, γ_{sl} and γ_{sa} are the liquid-air, solid-liquid and solid-air interfacial tensions respectively while area $\pi R'^2$ of the three-phase contact line is $\pi R_p^2 \sin^2\theta = R_p^2(1 - \cos^2\theta)$ and thus

$$A_{sa} = 2\pi R_p d = 2\pi R_p^2(1 - \cos\theta) \tag{4.4}$$

and

$$A_{sl} = 2\pi R_p d' = 2\pi R_p^2 (1 + \cos\theta) \tag{4.5}$$

If the particle is detached and is fully immersed in the liquid drop as illustrated in Figure 4.1d, its surface free energy will become

$$G_2 = \gamma_{la} A_{la} + \gamma_{sl} A_s = \gamma_{la} A_{la} + \gamma_{sl}(A_{sa} + A_{sl}) \tag{4.6}$$

The A_s is the particle surface area which is equal to the sum of A_{sl} and A_{sa}. The free energy ΔG for detaching the particle into the liquid phase, Equation (4.7), is obtained by subtracting Equation (4.3) from (4.6).

$$\Delta G = \pi R'^2 \gamma_{la} - (\gamma_{sa} - \gamma_{sl}) A_{sa} \tag{4.7}$$

When combined with the radius of the three-phase contact line and Equations (4.4), (4.7) can be written as

$$\Delta G = \pi R_p^2 \gamma_{la}(1 - \cos^2\theta) - 2\pi R_p^2 (1 - \cos\theta)(\gamma_{sa} - \gamma_{sl}) \tag{4.8}$$

Similarly, by subtracting Equation (4.3) from the free energy G_3 of detaching the particle and immersing it fully in air (Figure 4.1d), Equation (4.9), the free energy $\Delta G'$ for detaching the particle into the air phase is obtained as given in Equation (4.10), which can also be written as in (4.11).

$$G_3 = \gamma_{la} A_{la} + \gamma_{sa} A_s = \gamma_{la} A_{la} + \gamma_{sa}(A_{sa} + A_{sl}) \tag{4.9}$$

$$\Delta G' = \pi R'^2 \gamma_{la} + (\gamma_{sa} - \gamma_{sl}) A_{sl} \tag{4.10}$$

$$\Rightarrow \Delta G' = \pi R_p^2 \gamma_{la}(1 - \cos^2\theta) + 2\pi R_p^2 (1 + \cos\theta)(\gamma_{sa} - \gamma_{sl}) \tag{4.11}$$

Based on Young's law, $\gamma_{la}\cos\theta = \gamma_{sa} - \gamma_{sl}$, and rearrangement, Equations (4.8) and (4.11) can be written respectively as

$$\Delta G = \pi R_p^2 \gamma_{la}(1 - \cos\theta)^2 \tag{4.12}$$

$$\Delta G' = \pi R_p^2 \gamma_{la}(1 + \cos\theta)^2 \tag{4.13}$$

Equations (4.12) and (4.13) are generally written as

$$\Delta G = \pi R_p^2 \gamma_{la}(1 - |\cos\theta|)^2 \tag{4.14}$$

The energy released from particle attachment, bearing in mind that the particles come mainly from the air phase, is generally written as

$$\Delta G = -\pi R_p^2 \gamma_{la}(1 - |\cos\theta|)^2 \tag{4.15}$$

Equation (4.15) is obtained by subtracting (4.9) from (4.3) followed by rearrangement. For a particle coming from the liquid phase, the analogues of Equation (4.15) are obtained by subtracting Equation (4.6) from (4.3), followed by rearrangement.

Example 4.2: Calculating the Bond Number and Free Energy of Particle Attachment and Minimum Energy for Particle Detachment

The equilibrium contact angle of a particle (diameter ~6 μm), adsorbing from air onto the surfaces of a water drop (30 μL) is ~110°. If the water drop is at 25°C where the surface tension is ~72 mN m^{-1}, calculate the Bond number and the energy (in $k_B T$) released from the process and the minimum energy required to detach the particle into the air or water phase. The k_B is the Boltzmann's constant (~1.381 × 10^{-23} J K^{-1}), and T is the absolute temperature. Take the density of air and water (at 25°C) as 1.184 kg m^{-3} and 997 kg m^{-3}, respectively, and gravitational acceleration as 9.8 m s^{-2}.

Method

Substitute $R_p = 3 \times 10^{-6}$ m, $\gamma_{la} = 72 \times 10^{-3}$ N m^{-1} and $g = 9.8$ m s^{-2} into Equation (4.2) and calculate Bo at once for the first part. For the second part substitute the quantities given into Equations (4.15), (4.13) and (4.12) to calculate the energy released from particle attachment and the minimum energy for detaching the particle and immersing it fully in the air and water phases, respectively. Finally, write the energy in terms of the thermal energy ($k_B T$) in each case.

Answer

$$\text{Bo} = \frac{\Delta\rho g R_p^2}{\gamma_{la}} = \frac{(997-1.184)\ \text{kg m}^{-3} \times 9.8\ \text{m s}^{-2} \times (3\times10^{-6})^2\ \text{m}^2}{\underbrace{72\times10^{-3}\ \text{kg m s}^{-2}\ \text{m}^{-1}}_{N}} = 1.2\times10^{-6}$$

For energy released from particle attachment:

$$\Delta G = -\pi R_p^2 \gamma_{la}(1 - |\cos\theta|)^2$$

$$= -\pi \times (3\times10^{-6})^2\ \text{m}^2 \times 72\times10^{-3}\ \text{N m}^{-1} \times (1-|\cos 110°|)^2 = -8.81\times10^{-13}\ \text{J}$$

At 25°C, 1 $k_B T$ equals 4.11 × 10^{-21} J and therefore ΔG (in $k_B T$) is −2.1 × 10^8 $k_B T$.
　　The minimum energy for detaching the particle and immersing it fully in the air phase is

$$\Delta G = \pi R_p^2 \gamma_{la}(1 + \cos\theta)^2$$

$$= \pi \times (3\times10^{-6})^2\ \text{m}^2 \times 72\times10^{-3}\ \text{N m}^{-1} \times (1 + \cos 110°)^2 = 8.81\times10^{-13}\ \text{J}$$

$$= 2.1 \times 10^8\ k_B T$$

The minimum energy for detaching the particle and immersing it fully in the water phase is

$$\Delta G = \pi R_p^2 \gamma_{la}(1 - \cos\theta)^2$$

$$= \pi \times (3 \times 10^{-6})^2 \text{ m}^2 \times 72 \times 10^{-3} \text{ N m}^{-1} \times (1 - \cos 110°)^2 = 3.67 \times 10^{-12} \text{ J}$$

$$= 8.9 \times 10^8 \ k_B T$$

Example 4.2 shows that the energy released from particle adsorption is large and negative while the minimum amount of energy required for detaching the particle and immersing it fully in the air or water phase is large and positive. On this basis, particle attachment is considered favorable and spontaneous while particle detachment is considered unfavorable and difficult. In fact, particles are considered irreversibly adsorbed at the drop surfaces as a result.

4.2 PREPARATION OF LIQUID MARBLES

Liquid marbles are prepared in two ways. In the first, known as the "preformed drop template" method, a liquid drop, of suitable size, is rolled on a powdered bed of the particles prepared on a suitable substrate (Aussillous and Quéré 2006). The particles encapsulate the liquid drop by forming a monolayer or multilayers around it. During liquid marble formation, a drop may or may not become completely coated with the particles. It has been reported (McHale *et al.* 2007) that simply depositing a liquid drop onto a loose bed of the particles results in the underside becoming coated and particles lifting from the particle bed. This is as a result of the drop attempting to minimize its surface area towards that of a sphere and the lack of any significant particle resistance other than gravity as the particles are not bound to each other. Subsequent evaporation of such a system results in skirt of particles appearing to rise around the drop, creating a self-coating effect. This is presumably due to conservation of the surface area coverage with reducing drop volume *via* evaporative, capillary, or other driven bulk surface flow. If the initial deposition of a drop results in a significant impact, it will collect particles during the spreading stage of the impact and the coating will remain and compress as the drop contact line recedes or as the drop rebounds. This has been studied with water and glycerol drops using polytetrafluoroethylene (PTFE) powdered particles (size 1–100 μm) (Eshtiaghi *et al.* 2009). It was found that the extent of drop coverage is related to the initial kinetic energy of impact (Eshtiaghi *et al.* 2009). It has also been shown that if the particles are large, they may not be able to climb up (or remain on top) of the drop and a liquid marble may appear open at the top (McEleney *et al.* 2009). To form a liquid marble, the drop diameter d_d is expected to be much greater than the primary particle diameter d_p so that the particles can spread around it. Quantitatively, a marble forms when $d_d > 25 \ d_p$ (Eshtiaghi and Hapgood 2012). The general framework for liquid marble formation *via* the preformed drop template regime is given in Figure 4.2.

A distorted liquid pool or puddle forms when the gravity forces dominate and the Bond number Bo is greater than 1. This is different in the case of quasi-spherical marbles where the surface tension forces dominate and Bo is less than 1. The Bo for a liquid drop in air is

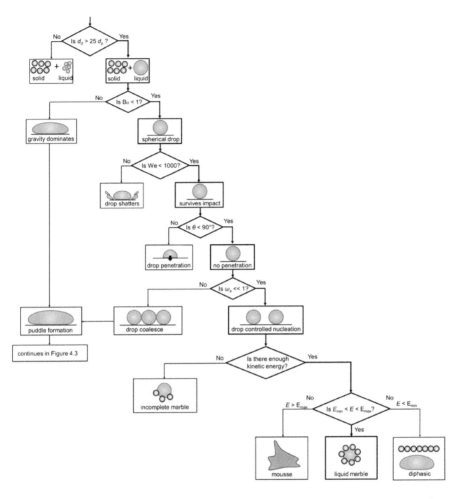

FIGURE 4.2 Framework for liquid marble formation *via* the preformed drop template method. (From Eshtiaghi, N. and Hapgood, K.P., *Powder Technol.*, 223, 65–76, 2012.)

$$Bo = \frac{\Delta \rho g R_d^2}{\gamma_{la}} \qquad (4.16)$$

in which $\Delta \rho$ is density of liquid minus density of air and R_d is the radius of the liquid drop. The drop is expected to remain intact and not shatter into smaller drops upon impacting the particle bed. This is predicted using the dimensionless Weber We and Ohnesorge Oh numbers, which describe the tendency for a liquid drop to either stay intact or split. The We compares the deforming inertial forces with the stabilizing cohesive forces (expressed in terms of the liquid surface tension) while the Oh compares the viscous forces with the inertial and cohesive forces. The We, Equation (4.17)

$$We = \frac{\rho d_d v^2}{\gamma_{la}} \tag{4.17}$$

is the ratio of the inertial forces to the cohesive forces for a liquid drop moving through air or bulk liquid phase. The ρ is the liquid density and v is the velocity of the drop in air. The We indicates whether the kinetic energy is dominant over the cohesive energy. The modified Weber number We^* [Equation (4.18)] is the ratio of the kinetic energy E_k on impact to the surface energy E_s.

$$We^* = \frac{E_k}{E_s} = \frac{\pi \rho d_d^3 v^2 / 12}{\pi d_d^2 \gamma_{la}} = \frac{We}{12} \tag{4.18}$$

The Oh, Equation (4.19) where η is the liquid viscosity, is the ratio of the viscous forces to the square-root of the inertial and surface tension forces.

$$Oh = \frac{\text{viscous forces}}{\sqrt{\text{inertial force} \times \text{surface tension force}}} = \frac{\eta}{\sqrt{\rho \gamma_{la} d_d}} \tag{4.19}$$

It has been suggested that liquid drops remain intact when We < 1000 and Oh > 0.05 (Agland and Iveson 1999). However, the additional condition of Oh > 0.05 is inconsistent with experimental data on liquid marble formation and can be ignored (Eshtiaghi and Hapgood 2012). Bulk liquid marble formation is possible with the preformed liquid drop template method if an appropriate spraying device is used. In order to obtain small individual liquid drop templates, the dimensionless spraying flux ω_a is expected to be low (*i.e.* $\omega_a < 0.1$) to prevent drop coalescence. Additionally, Forny *et al.* (2009) have shown that the energy E applied per unit mass of powder is key during liquid marble formation. The energy is expected to lie between the minimum threshold energy E_{min} and the maximum threshold energy E_{max}. When $E < E_{min}$, the liquid drop and particles remain separated and no marble is formed. On the contrary, a mousse or foam forms when $E > E_{max}$.

Example 4.3: Predicting Whether a Liquid Drop Will Remain Intact or Split upon Impacting a Particle Bed

A spherical water drop (40 µL) is released from a height (1 cm), with velocity 1 m s⁻¹, onto a loosely packed powdered bed of particles (diameter about 10 µm). If the water drop is at 25°C where its surface tension and density are about 72 mN m⁻¹ and 997 kg m⁻³, respectively, say whether it will remain intact or split upon impacting the powdered particle bed.

Method

The appropriate mathematical formula to use is given in Equation (4.17). Substitute ρ = 997 kg m⁻³, v = 1 m s⁻¹, γ_{la} = 72 × 10⁻³ N m⁻¹ and $d_d = \sqrt[3]{32 \text{ Volume of drop}/3\pi} = \sqrt[3]{32 \times 40 \times 10^{-9} \text{ m}^3/3\pi} = 5.1 \times 10^{-3}$ m into the formula to find We at once. The water drop will remain intact if We < 1000.

Answer

$$We = \frac{\rho d_d v^2}{\gamma_{la}} = \frac{997 \text{ kg m}^{-3} \times 5.1 \times 10^{-3} \text{ m} \times 1^2 \text{ m}^2 \text{ s}^{-2}}{72 \times 10^{-3} \text{ N m}^{-1}} = 70.6$$

Because We < 1000, the drop will remain intact upon impacting the powdered particle bed.

The second method of marble preparation is known as the "mechanical dispersion" method and involves agitating a mixture of the particles and the liquid ("bulky puddle"). The liquid breaks into smaller drops and become coated with particle layer(s), provided $\omega_a \gg 1$ and $\theta > 110°$ (Aussillous and Quéré 2006, Binks *et al.* 2014). The contact angle requirement is necessary as the highly energetic agitation conditions can lead to the immersion of moderate 'phobic particles into the liquid phase. Depending on the degree of agitation, marbles of any liquid can be prepared. It must be noted that relatively high tension liquids like glycerol and water require a relatively higher degree of agitation compared with relatively low tension ones like oils. If the degree of agitation is low, the liquid and the powder remain as two different phases (*i.e.* biphasic). A minimum threshold of agitation is required for effective dispersion of the liquid through the powder and formation of liquid marbles. Excessive agitation causes the liquid marbles to collapse into a foam or mousse. As mentioned earlier, successful liquid marble formation requires the energy supplied per unit mass (E), in the form of agitation, to lie between the minimum (E_{min}) and the maximum (E_{max}), irrespective of the liquid to particle mass ratio. The framework for the formation of liquid marbles *via* this method is summarized in Figure 4.3.

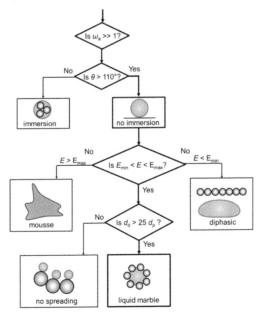

FIGURE 4.3 Framework for liquid marble formation *via* the mechanical dispersion method. (From Eshtiaghi, N. and Hapgood, K.P., *Powder Technol.*, 223, 65–76, 2012.)

Large-scale production of liquid marbles, *e.g.* in the production of dry powdered liquids (*i.e.* dry water and oils), which are a collection of plenty tiny liquid marbles, has been achieved using this method (Binks and Murakami 2006, Binks *et al.* 2014, Tyowua 2014). The downside of this method is the difficulty associated with tuning the size of the individual liquid marbles in the powder.

Different powdered particles are used in the preparation and stabilization of liquid marbles. The particles are expected to have low affinity for the surfaces of the liquid drop so that they will not be completely wetted by the liquid drop. The examples of these particles are given in Table 4.1 and are discussed here. McEleney *et al.* (2009) used water and hydrophobic copper particles of different sizes (9, 20 and 320 μm) to prepare liquid marbles using petri dishes and hemispherical clock glass. These hydrophobic particles (contact angle with water approaching 160°) were prepared *via* silver deposition technique. Hydrophobic poly-methylmethacralate (PMMA) powdered particles with particle size 42 μm and contact angle of 120° were further used to determine the effect of powder density on marble stability. They argued that successful formation of a liquid marble depends on the type of powdered particles, hydrophobicity, density, particle size, shape, and liquid marble formation technique. Stimulus-responsive liquid marbles have also been prepared by using pH-responsive hydrophobic polyacid-stabilized polystyrene latex particles of average diameter ~80 nm to encapsulate water drops of pH 2.5 (Dupin *et al.* 2011). The sterically-stabilized polystyrene latex particles were prepared with a poly-acid macromonomer. These liquid marbles remained intact when placed on the surfaces of liquid water adjusted to pH 4 or below, but addition of a base to the aqueous solution caused imme-diate destruction of the marble because of the hydrophilic nature of the stabilizer chains above the pH. This further suggests that the stability of these liquid marbles depends on pH. Millimetre- and centimetre-sized liquid marbles have also been prepared by rolling between 15 μL and 2 mL volumes of water, aqueous gum, and glycerol solu-tions on dry submicrometer-sized and sterically stabilized pH responsive polystyrene (PS) latex particles (Fujii *et al.* 2011). The particles were synthesized by dispersion polymerization in isopropyl alcohol with poly [2-(diethylamino) ethyl methacrylate] (PDEA) based on microinitiator. Evidence from scanning electron microscope (SEM) and fluorescence microscope indicate that the stability of the marbles resulted from adsorption of flocs of the PDEA-PS particles on the surfaces of the water drops. The liquid marbles were stable for about 1 hour on the surfaces of an aqueous solution whose pH was above 8 but disintegrated immediately in the presence of a lower one. Mechanically robust magnetic liquid marbles, prepared by rolling a water drop on highly hydrophobic Fe_3O_4 nanoparticles (size 8.6 nm) have been reported (Bhosale *et al.* 2008). The particles were synthesized by co-precipitation of Fe^{2+} and Fe^{3+} salts in an ethanol-water solution with ammonia in the presence of a fluorinated alkyl silane which hydrolyzed in solution to form a coating on the Fe_2O_3 nanoparticles. The mag-netic liquid marbles have a remarkable ability to be opened and closed reversibly under the action of a magnetic field. Based on this ability, adjusting the liquid inside the marble drop and coalescing of two liquid marbles into larger ones were dem-onstrated. Dupin *et al.* (2009) have reported the preparation of stimulus-responsive liquid marbles by rolling a 10 μL of deionized water on hydrophobic PDEA-PS latex particles. The liquid marbles remained intact after transfer onto a glass slide or onto

TABLE 4.1
Summary of Liquid, Powdered Particles, Particle Diameter, and Three-Phase Contact Angle θ from Scattered Literature on Liquid Marbles. The Number in Brackets after a Liquid Represents its Surface Tension (mN m⁻¹) at 25°C

Liquid	Powdered Particles	Diameter/μm	θ/°	References
Water (71.8)	Hydrophobized lycopodium	–	–	Aussillous and Quéré (2001)
	Sporopollenin capsules	25	–	Binks *et al.* (2011)
	Hydrophobized Cu	9–320	157	McEleney *et al.* (2009)
	PMMA	42	120	McEleney *et al.* (2009)
	PTFE	5–6	97–102	Tosun and Erbil (2009)
	Graphite	2–30	155.4	Doganci *et al.* (2011)
	Hydrophobic silica of residual SiOH 61%–14%	0.02–0.03	73–126	Zang *et al.* (2014)
	Hydrophobized spherical glass beads	42–165	110	Whitby *et al.* (2012)
	Janus particles composed of silica and α-Fe_2O_3	~0.395	72 & 100	Kim *et al.* (2010)
Water, glycerol (64.0),	Hydrophobized glass beads,	65–191	–	Eshtiaghi and Hapgood (2012)
water-glycerol mixtures	PTFE, fumed silica Aerosil R202	1–100, –0.016	–	Eshtiaghi and Hapgood (2012)
Water, dimethyl sulfoxide DMSO (43.5), toluene (28.4), hexadecane (26.3), ethanol (22.1) and octane (21.6)	Fluorinated decyl polyhedral oligomeric silsesquioxane POSS/ Fe_2O_3 composite	<70	143–171	Xue *et al.* (2010)
Water, diiodomethane (58), DMSO, dimethyl formamide (36.4), 1,4-dioxane (33.7),	Poly[2-(perfluorooctyl) ethylacylate] PFA-C_8	1.4 ± 0.3	–	Matsukuma *et al.* (2011)

(Continued)

TABLE 4.1 (*Continued*)
**Summary of Liquid, Powdered Particles, Particle Diameter, and
Three-Phase Contact Angle θ from Scattered Literature on Liquid
Marbles. The Number in Brackets after a Liquid Represents its Surface
Tension (mN m⁻¹) at 25°C**

Liquid	Powdered Particles	Diameter/μm	θ/°	References
toluene, ethanol and methanol (22.5)				
Water and oils (26–63)	Fluorinated silica	0.02–0.03	28–149	Binks and Tyowua (2013)
Water and oils (20–37)	Fluorinated platelet clay and	–	39–148	Binks *et al.* (2015)
	ZnO particles			Binks *et al.* (2014)
Galinstan metal	PTFE, silica, WO₃, TiO₂, MoO₃, In₂O₃ and carbon nanotubes	–	–	Sivan *et al.* (2013)

the surfaces of liquid water. Partial buckling and collapse of the marbles gradually occurred after several hours at 20°C due to slow evaporation of water in the marble core. However, their stability was enhanced by addition of electrolytes in the drop. The non-volatile solute reduced the rate of evaporation from the encapsulated drops. Preparation of aqueous liquid marbles using very hydrophobic particles such as PTFE (Eshtiaghi *et al.* 2009, Bormashenko *et al.* 2009a), natural lycopodium (Bormashenko *et al.* 2009a, 2009b), hydrophobized lycopodium (McHale *et al.* 2008), polyethylene (PE) (Bormashenko *et al.* 2009a), hydrophobic silica powder (Bhosale *et al.* 2008, Bhosale and Panchagnula 2010), and other particles like carbon black (Bormashenko *et al.* 2010a), graphite (Dandan and Erbil 2009), polyvinylidene fluoride (PVDF) (Bormashenko *et al.* 2008, 2009a), and Janus particles (Kim *et al.* 2010) have also been reported. Using relatively low surface energy fluorinated fumed silica particles (Binks and Tyowua 2013), fluorinated clay particles (Binks *et al.* 2014, 2015), and ZnO particles (Binks *et al.* 2015), oil liquid marbles were prepared by rolling a drop of the oil on a bed of the particles on a Teflon substrate. The oil liquid marbles remain stable to evaporation for more than three days.

4.3 PROPERTIES OF LIQUID MARBLES

4.3.1 STATICS OF LIQUID MARBLES

Liquid marbles are coated with particles of relatively low surface energy. This changes their interaction with a solid substrate from a liquid-solid interaction to a solid-solid interaction. Many of the observed static properties of liquid marbles result from the solid-solid interaction, which in turn depends on the degree of protrusion of the particles out of the drop surfaces into the air phase. The greater the degree of particle protrusion, the larger the separation of the encapsulated liquid drop from a substrate (Bormashenko *et al.* 2009d). Since there is no contact between the encapsulated liquid and the solid substrate upon which a liquid marble rests, stable liquid marbles are completely non-wetting and exhibit a large apparent contact angle θ_A (> 115°, measured into the liquid) (Binks and Tyowua 2013) with the substrate as shown in Figure 4.4a–c.

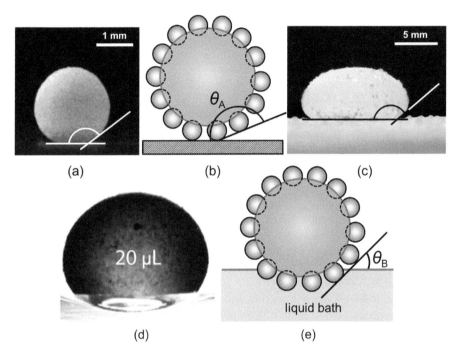

(a)   (b)   (c)

(d)   (e)

FIGURE 4.4 Photograph of a quasi-spherical (10 μL) aqueous liquid marble (a), and its schematic illustration (b), making an apparent contact angle θ_A of ~135° with the substrate. Photograph of an aqueous puddle (500 μL) making θ_A of ~135° with the substrate (c). Both marbles are stabilized by PVDF powdered particles. (From Bormashenko, E. *et al., Colloids Surf. A*, 351, 78–82, 2009.) (d) An aqueous liquid marble (20 μL) resting on a water bath at ambient temperature and (e) its schematic illustration. (From Vadivelu, R.K. *et al., Sci. Rep.*, 5, 15083, 2015.)

The θ_A also depends on the degree of particle protrusion from the drop surfaces into the air phase with large particle protrusion giving rise to a large value of θ_A and vice versa. Contact angle hysteresis is negligibly small in liquid marbles, compare to bare liquid drops where it is significant, due to the non-contraction of the particle layers encapsulating the liquid drop when its volume decreases. Liquid marbles are also able to rest on other liquids (Figure 4.4d) provided the liquid does not wet down completely the particle length separating the encapsulated liquid from the supporting liquid bath. The floating of liquid marbles on the surfaces of liquids like water has been reported by many researchers (Feng *et al.* 2002, Verplanck *et al.* 2007, Gao *et al.* 2009). The lack of direct contact between the encapsulated liquid and the solid substrate or liquid has been confirmed by the absence of any colored trace for NaOH containing liquid marbles rolling across a phenolphthalein coated surface (Bormashenko *et al.* 2009d). When a liquid marble rests on a liquid bath with a contact angle θ_B with the particles encapsulating the liquid marble (Figure 4.4e), the separation d_B between the liquid bath and the encapsulated liquid within the marble is $-R_p(\cos\theta + \cos\theta_B)$. For an aqueous liquid marble stabilized by PTFE particles resting on a water bath, $d_B \sim R_p$ (Bormashenko *et al.* 2009d).

When at rest on a substrate, the shape of liquid marbles is dictated by the interplay of the surface tension of the encapsulated liquid and gravity. While the surface tension forces dominate in small liquid marbles and are quasi-spherical in shape with a small flat spot where they rest on a solid substrate, large ones are flattened by gravity forces and have a puddle shape (Bormashenko *et al.* 2009d) as shown in Figure 4.4a–c. The radius δ_m of the flat spot contact zone depends on the volume of the encapsulated liquid. For aqueous liquid marbles, the size of the flat spot follows the rule (Aussillous and Quéré 2001)

$$\frac{\delta_m}{\kappa^{-1}} = \sqrt{\frac{2}{3}}\left(\frac{R_m}{\kappa^{-1}}\right)^n \tag{4.20}$$

The R_m is the radius of the marble before resting on the substrate, $n = 3/2$ (for large marbles, *i.e.* >100 μL) or $n = 2$ (for small marbles, *i.e.* <100 μL), κ^{-1} is the capillary length given by $(\gamma_{la}/\rho g)^{0.5}$ with ρ and g being the density of the liquid and the acceleration due to gravity respectively. For quasi-spherical liquid marbles, $R_m \ll \kappa^{-1}$, but for puddle-shaped ones $R_m \gg \kappa^{-1}$ (Aussillous and Quéré 2006). For example, the κ^{-1} of water at 25°C is 2.7 mm and small aqueous liquid marbles (<100 μL and $R_m < 2.9$ mm) stabilized by poly(2-(diethylamino)ethyl methacrylate)-modified polystyrene latex particles are quasi-spherical (Fujii *et al.* 2011). However, larger ones (>100 μL and $R_m > 2.9$ mm) adopt a puddle shape (Fujii *et al.* 2011). The shape of liquid marbles is also related to its Bo defined mathematically as $\Delta\rho g R_m/\gamma_{la}$. It has been reported that Bo << 1 gives quasi-spherical marbles while Bo >> 1 gives puddle-shaped ones (Aussillous and Quéré 2006). These conditions correspond to those given above in terms of R_m and κ^{-1}. The δ_m generally varies inversely with the κ^{-1} as shown in Figure 4.5a. However, the θ_A of liquid marbles increases as the κ^{-1} increases (Figure 4.5b). The apparent contact angle of a marble on a substrate is usually greater than the corresponding equilibrium contact angle of the "bare" liquid on the same substrate as shown in the inset of Figure 4.5b.

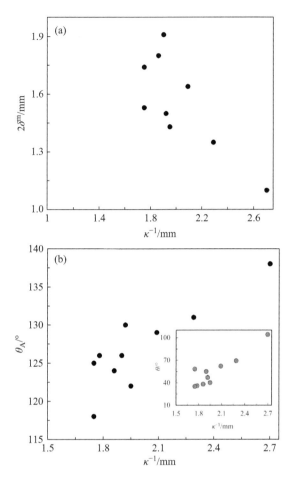

FIGURE 4.5 (a) The diameter ($2\delta_m$) of contact zone and (b) the apparent contact angle (θ_A) of different liquid marbles (10 μL) on Teflon *versus* the capillary length κ^{-1} (20°C) of the encapsulated liquids: (a) shows a general decrease in $2\delta_m$ as the κ^{-1} increases while (b) shows an increase in θ_A as κ^{-1} increases. The inset in (b) is the corresponding equilibrium contact angle θ of the "bare" liquids on Teflon *versus* the κ^{-1} of the liquids. (From Binks, B.P. and Tyowua, A.T., *Soft Matter*, 9, 834–845, 2013.)

Example 4.4: Variation of the Radius of the Flat Spot Contact Zone with the Volume of the Encapsulated Liquid

For water drops (5–100 μL), show that the radius δ_m of the flat spot contact zone depends on the volume of the encapsulated drops for a temperature of 25°C.

Method

Using Equation (4.20), calculate δ_m for water marbles (5–100 μL) and plot it against the volume of the encapsulated drops. The plot shows that δ_m increases with the volume of the marbles.

Answer

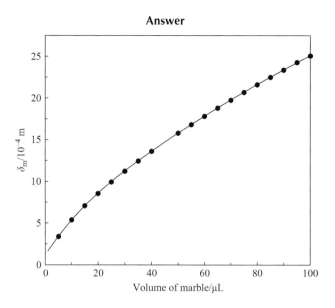

4.3.2   DYNAMICS OF LIQUID MARBLES

Due to the high contact angle between liquid marbles and the substrate surfaces and the low hysteresis associated with it, liquid marbles exhibit unique dynamic properties. For example, they move faster than the corresponding bare liquid drops. This is due to the absence (or very weak role) of contact lines, close to which most of the viscous dissipation occurs. The motion of liquid marbles will be discussed under three themes, namely creeping and quick motions and bouncing.

4.3.2.1   Creeping Motions

Suppose the encapsulated liquid is of high viscosity so that the main source of friction is viscosity (even as the contact lines are absent) and it is placed on a plane inclined at an angle α and the force driving motion is gravity, the motion of puddles and quasi-spherical liquid marbles can be distinguished.

4.3.2.1.1   Puddle Liquid Marbles

An object placed on an inclined surface would either move down by rolling or sliding. It has been shown that liquid marbles placed on an inclined substrate move down by only rolling (Mahadevan and Pomeau 1999, Richard and Quéré 1999). Generally, small marbles move down inclined planes faster than large ones (*i.e.* puddles) (Aussillous and Quéré 2001, 2006, Pike *et al.* 2002). The dynamics of liquid marbles is controlled by the capillary number *Ca*, which represents a balance between the action of viscous forces and surface tension forces on the drop. The *Ca* is defined as $\eta V_o/\gamma_{la}$, where η is the viscosity of the encapsulated liquid and V_o is the (terminal) velocity of the liquid marble. It was reported (Richard and Quéré 1999, Bormashenko *et al.* 2010b) that for aqueous liquid marbles (10 μL) rolling down an inclined substrate with a V_o of ~0.1 ms^{-1}, *Ca* << 1. Nonetheless, for glycerol liquid marbles of the same volume rolling down the same substrate with the same velocity, $Ca \sim 1$. This means viscous dissipation is negligible in the aqueous liquid marbles compared with the viscous glycerol ones where it is significant. The viscous dissipation would tend to slow down the glycerol marbles compared with the aqueous ones.

The V_o is obtained from the balance between viscous friction $\eta A'V/h$ (or resistance) and the force of gravity $mg \sin \alpha$ acting on the marble. The A' is the area of contact between the puddle and the plane, V is the velocity of the puddle and h is the puddle height $\sim 2\kappa^{-1}$. The viscous friction tends to hold the marble back on the plane while the gravitational force tends to drive it down the plane. Substituting the appropriate expressions for A' and h and noting that for very small angles of inclination, $i.e.$ α (in rad) $\ll 1$, $\sin \alpha \sim \alpha$, followed by rearrangement gives Equation (4.21) which translates to (4.22).

$$\frac{1}{4} \times \eta \frac{V_o}{\kappa^{-2}} \times \text{puddle volume} = \rho g \alpha \times \text{puddle volume} \tag{4.21}$$

$$\Rightarrow \eta \frac{V_o}{\kappa^{-2}} \sim \rho g \alpha \tag{4.22}$$

Upon considering κ^{-2} as $\gamma_{la}/\rho g$, Equation (4.22) becomes (4.23).

$$V_o \sim \frac{\gamma_{la}}{\eta} \alpha \tag{4.23}$$

The numerical coefficient of Equation (4.23) lies between 1 and 4/3 and depends on the boundary condition at the top surface. For large (diameter between 1 and 2 cm) glycerol puddles, the coefficient is found to be larger (1.5 ± 0.5) than the expected value. This might be due to the finite size of the puddle. Equation (4.23) shows that V_o (i) varies directly with γ_{la} and α and inversely with η and (ii) is independent of the product ρg, which fixes both the driving gravity force and the resistive viscous friction. When the product ρg is large, the puddle height is small and the viscous friction is large too, thereby reducing the velocity. However, when the product ρg is small, the puddle height is large and the viscous friction is small. In this case the velocity is not significantly affected. This is in line with the work of Aussillous and Quéré (2006) where puddles of different viscosities, tuned by varying the fraction of glycerol and water in them, were placed on planes inclined at angles between $1°$ and $5°$. They reported that the velocity of the puddles was independent of their size but decreases with increasing viscosity and decreasing angle of inclination. The relationship between the motion of a particle grain on a puddle marble and its motion was also investigated. Using glycerol coated with about a monolayer of lycopodium and an inclination angle of $2°$, it was observed that the velocity of a particle grain V_{grain} used as a marker was about twice the velocity of the puddle itself ($i.e.$ $V_{grain} = 1.85 V_o$) while the motion of the puddle was close to the caterpillar motion (Aussillous and Quéré 2006). The value of 1.85 indicates an intermediate behavior between a free surface (coefficient 1.5) and a caterpillar motion (coefficient 2). It is assumed that the solid particles can move and rearrange to support a change of speed between the top and the bottom of the drop. Equation (4.23) predicts the terminal velocity of puddles quite well. However, deviations occur when the $\eta < 100$ mPa s and when the air friction dominates over viscous friction.

4.3.2.1.2 Small Liquid Marbles

Small, quasi-spherical marbles placed on inclined planes roll down faster than their puddle counterparts. The smaller the liquid marble, the faster it rolls down the plane. These marbles are more mobile compared with normal uncoated liquid drops in air. Provided the Reynolds number Re [the ratio of inertial forces to viscous forces, Equation (4.24),

$$\text{Re} = \frac{d_m \rho V}{\eta} \tag{4.24}$$

where V is the marble velocity and d_m is the marble diameter] is small, the motion of viscous quasi-spherical liquid marbles is considered as a superposition of solid-like rotation (producing no dissipation) due to the presence of the encapsulating solid particles with viscous friction (producing dissipation) localized in the contact zone. This behavior was first reported by Mahadevan and Pomeau (1999) and it is said to occur in the Mahadevan-Pomeau regime. At relatively small velocities, marble deformations are insignificant, and the size of the contact zone remains as given in Equation (4.20), and the velocity of the marble can be obtained by balancing the gain of gravitational energy $mgV \sin \alpha$ with the viscous dissipation per unit time $\eta \int_\Omega (\nabla u)\, d\Omega$ (Mahadevan and Pomeau 1999). The Ω is the volume (of order δ_m^3) over which viscous dissipation takes place and u is the velocity field in the marble. The velocity gradients in the contact zone are of the order V/R_m. This yields a dissipation energy which scales as $\eta(V/R_m)^2 \delta_m^3$. This energy balances with the gravitational energy per unit time to give the terminal velocity V', Equations (4.25) to (4.28), of the marble, with a numerical coefficient of 1 as confirmed by Aussillous and Quéré (2006).

$$V' \sim \frac{\rho g R_m^5 \sin \alpha}{\eta \delta_m^3} \tag{4.25}$$

$$\Rightarrow V' \sim \frac{\rho g R_m^5 \alpha}{\eta \delta_m^3} \text{ (at low } \alpha) \tag{4.26}$$

$$\Rightarrow V' \sim \frac{\rho g R_m^5 \alpha}{\eta \left(\frac{2}{3}\right)^{3/2} R_m^6 \kappa^3} \sim \frac{\gamma_{la}}{\eta} \alpha \times \frac{1}{(2/3)^{3/2} R_m \kappa} \tag{4.27}$$

$$\Rightarrow V' \sim V_o \frac{\kappa^{-1}}{R_m} \tag{4.28}$$

Viscous globules moving under gravity generally obey Stokes law, *i.e.* their velocity increases strongly with the globule size (as square of radius). This is different with viscous non-wetting drops like small liquid marbles, for which their velocity increases as their size decreases in line with Equation (4.28), which is particularly due to the quadratic dependence of the size of the contact zone with the marble radius as given in Equation (4.20). This behavior is a unique characteristic of the

Mahadevan-Pomeau regime, but there are limitations however. The first limitation is related to the liquid viscosity. The velocity of marbles of relatively small viscosity ($\eta < 100$ mPa s) increases with their size violating Equation (4.28). Similarly, the law for viscous puddles, Equation (4.23), is also not obeyed for $\eta < 10$ mPa s. The second limitation concerns the angle of inclination. It has been reported (Aussillous and Quéré 2006) that viscous marbles (puddles and drops), for which $\eta = 1000$ mPa s, running on planes inclined at $\alpha > 10°$ exhibit velocities that are much larger than expected from Equations (4.23) and (4.28). The marbles are also reported to deform significantly, resembling a peanut or a doughnut, because of the centrifugation associated with the high rotational speed.

The condition on the viscosity and the inclination angle in the Mahadevan-Pomeau regime, where the rotation of small spherical drops is considered solid-like owing to a small Re, $i.e.$ Re > 1, is

$$\eta^2 > \rho\gamma_{la}\kappa^{-1}\alpha \tag{4.29}$$

Equation (4.29) is obtained by combining Equation (4.28) and the condition Re $= \rho V' R_{m}/\eta < 1$ typical of Mahadevan-Pomeau regime. For $\alpha \sim 10°$, the condition suggests a liquid viscosity less than 100 mPa s. For glycerol, the condition should be obeyed irrespective of the value of α. Deviations, however, occur when α is greater than 10°. Another condition to be satisfied by drops in the Mahadevan-Pomeau regime is a quasi-static shape while in motion. Unfortunately, inertial (centrifugation) and viscous forces deform moving drops (Aussillous and Quéré 2001). Because surface tension forces tend to restore the spherical nature of a drop, the We and Ca are expected to be less than 1, $i.e.$

$$We = \frac{\rho V'^2 R_{m}}{\gamma_{la}} < 1 \text{ and } Ca = \frac{\eta V'}{\gamma_{la}} < 1 \tag{4.30}$$

[Recall that the We represents the balance between inertial forces and cohesive forces, expressed in terms of surface tension forces, for a moving drop while the Ca represents the balance between the action of viscous forces and surface tension forces on a drop.] The condition on the Ca is more restrictive compared to that on We and combining it with the Re condition and Equation (4.29) gives (4.31).

$$R_{m} > \alpha\kappa^{-1} \tag{4.31}$$

The condition in Equation (4.31) is easily attained when the angle of inclination is small ($\alpha \to 0$), but it is very restrictive as $\alpha \to 1$.

4.3.2.2   Quick Motion

Increasing the velocity of liquid marbles, $e.g.$ through increasing the inclination angle, leads to shape deformation and the breakdown of the Mahadevan-Pomeau picture. Fast moving liquid marbles adopt different shapes (Aussillous and Quéré 2004). As the speed of motion increases, centrifugal forces become larger than centripetal forces and the shape of a liquid marble changes from quasi-spherical or puddle to peanut or doughnut (Figure 4.6 upper image) (Quéré and Aussillous 2002, Aussillous and Quéré 2004).

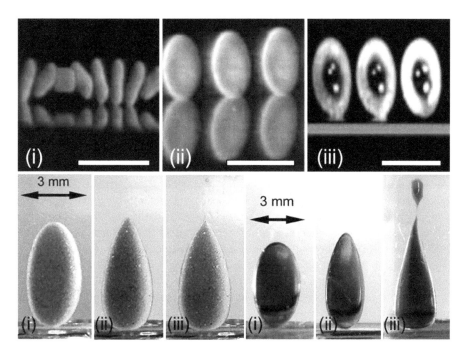

FIGURE 4.6 The upper image shows the glycerol liquid marbles stabilized by lycopodium grains rolling down a plate inclined at 35°. The initial radius R_m is (i) 1, (ii) 1.9 and (iii) 1.9 mm and the speed is ~1.2 m s⁻¹. The time intervals between the snapshots in (i)–(iii) are 1.1, 1.5 and about 3 ms respectively. The observed shapes are (i) peanut, (ii) disk, and (iii) doughnut or wheel shape. Scale bar = 1 cm. (From Aussillous, P. and Quéré, D., *J. Fluid Mech.*, 512, 133–151, 2004.) The lower image shows the lycopodium particle-stabilized water liquid marble (left) and a water liquid drop (right), both 20 µL, deformed in an electric field to a prolate-spheroid shape (i) followed by the formation of the Taylor cone (ii) and then jetting (iii) as the electric field strength increases. (From Bormashenko, E. *et al.*, *Colloid Polym. Sci.*, 291, 1535–1539, 2013a.)

Diamagnetically levitated liquid drops yield similar shapes (Hill and Eaves 2008). This observation disagrees with the ideas of Mahadevan and Pomeau (1999), who have argued that a non-wetting liquid drop should be in solid rotation, at small velocities or small Reynolds numbers, which minimizes viscous dissipation. The shapes mentioned above were seen at high speed and Re slightly greater than 1. As a consequence, the liquid drop is centrifuged and these shapes are observed. For leaving the spherical shape, the inertia must be greater than the capillary action which can be described in terms of the We. If We < 1, corresponding to velocities smaller than 30 cm s⁻¹ for millimetre-sized water or glycerol marbles, centrifugation hardly affects the shape of the marble whereas the reverse leads to the strong deformations reported above. The peanut state was initially predicted by Chandrasekhar (1965) and then described by Brown and Scriven (1980). It has also been observed in liquid drops rotating in space (Lee *et al.* 1998). In the peanut state, the two lobes remain connected because of the action of the substrate on which the peanut bounces (without breaking and leakage), bringing the lobes closer to each other. The doughnut state conserves the axisymmetry of the drop. As it accelerates, it gradually passes from a sphere to a disk and eventually to a doughnut (toroidal shape). The torus is not necessarily opened at the centre. This shape was initially proposed by Laplace and later by Rayleigh (1914).

The effect of an electric field on the shape of water liquid marbles (containing small amount of $KMnO_4$ for visualization) stabilized by three hydrophobic powdered particles, namely PVDF (diameter ~130 nm), PTFE (diameter ~1 μm), and lycopodium (diameter ~30 μm) has also been studied (Bormashenko *et al.* 2013a). This was compared with that of a bare water drop in the same electric field. A marble was immersed in polydimethylsiloxane ($\rho = 0.970$ g cm^{-3}, $\eta = 0.34$ Pa s and dielectric constant of 2.8) contained in a transparent rectangular container having a copper plate below (positive terminal) and above (negative terminal). The setup was connected and subjected to a uniform electric field (DC) (Bormashenko *et al.* 2013a). Instability of the water liquid marble shape followed by the appearance of a Taylor cone and jetting of a small drop was observed as shown in Figure 4.6 (lower image). This occurred at some critical value of the electric field. The square of the critical electric field depended linearly on the inverse of the radius of the marble. Apart from the lycopodium particle-stabilized liquid marbles and bare water drops, the extrapolation of the linear curve to the zero electric field yielded a finite value of the spherical marble radius. This indicates that the marbles are charged while the lycopodium particle-stabilized ones and the bare water drops are uncharged.

4.3.3 EFFECTIVE SURFACE TENSION OF LIQUID MARBLES AND ITS MEASUREMENT

The effective surface tension γ_{eff} of a liquid marble may be smaller or larger than the surface tension of the encapsulated liquid depending on the types of interactions (*e.g.* capillary and electrostatic) between the solid particles encapsulating the liquid drop (Bormashenko *et al.* 2009a, Arbatan and Shen 2011). The γ_{eff} of liquid marbles describes the surface tension of the liquid drop under the influence of the encapsulating powdered particles (Arbatan and Shen 2011). A simple explanation (see Chapter 1) of the origin of liquid surface tension is that the molecules at the liquid-air interface experience unbalanced attractive intermolecular forces resulting in a net pulling force toward the bulk of the liquid. For the shell of a liquid marble, the situation is more complicated because a substantial fraction of the liquid-air interface is covered by particles and becomes a liquid-solid interface (Nguyen *et al.* 2010). The powdered particles on the liquid surface may significantly change the curvature of the liquid drop surface at the microscopic level since the particles are not wetted by the liquid and are "floating" on the liquid surface, making micro indents on the liquid surface. The interactions between hydrophobic particles by capillary forces may affect the net balance of forces on the liquid marble surface. These forces may alter the surface tension of the liquid and cause the marble to have an "effective surface tension" that may or may not be the same as the surface tension of the encapsulating liquid (Aussillous and Quéré 2006). Only few literatures (Aussillous and Quéré 2006, McHale *et al.* 2008, Bormashenko *et al.* 2009a,b) have discussed the effective surface tension of liquid marbles. It has been shown that the γ_{eff} of marbles can be given by Equation (4.32) (Aussillous and Quéré 2004),

$$\gamma_{eff} = \gamma_{la} + \gamma_{int} \tag{4.32}$$

where the first term represents the surface tension of the liquid and the second one arises from all kinds of interactions (capillary, electrostatic, *etc.*) between the solid particles averaged over a unit area. A positive value of γ_{int} corresponds to a resultant attractive force while a negative one corresponds to a resultant repulsive force (Aussillous and Quéré 2004). This accounts for the reason why the γ_{eff} is either larger or smaller than the surface tension of the encapsulated liquid. It is worthy to mention that experimental data relating to the γ_{eff} of liquid marbles are scarce and experimental work devoted to the problem is still needed. Frequently used methods for establishing the γ_{eff} of liquid marbles include the puddle height method, analysis of marble shape, vibration of marbles, the Wilhelmy plate, and capillary rise methods.

4.3.3.1 The Puddle Height Method

The shape of a liquid marble is dictated by the balance of surface tension and gravity as mentioned previously. Liquid marbles are either quasi-spherical or puddle-shaped depending on their size, which in turn depends on the capillary length of the encapsulated liquid (Aussillous and Quéré 2006, McHale *et al.* 2008, Bormashenko *et al.* 2009a,b). For relatively large marbles or puddles, the thickness or height *h* is independent of its volume as shown in Figure 4.7 for different particle types. This is the basis of the puddle height method of effective surface tension measurement. The *h* can be expressed as

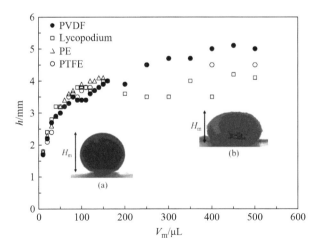

FIGURE 4.7 Height *h* of water liquid marbles *versus* volume V_m of the marbles. Inset: (a) quasi-spherical and (b) puddle-shaped marbles. Initially, *h* is dependent on V_m (a) and then independent of V_m (b). (From Bormashenko, E. *et al.*, *Langmuir*, 25, 1893–1896, 2009a.)

$$h = 2\kappa^{-1} \sin\left(\frac{\theta_A}{2}\right) \tag{4.33}$$

Suppose that for a liquid marble, the apparent contact angle θ_A is 180°, the maximal height h of the puddle will change asymptotically to twice the capillary length (Newton *et al.* 2007, Bormashenko *et al.* 2009a) and occurs at a radius of ~3.2κ^{-1} (Aussillous and Quéré 2006), although the maximal thickness of a drop is 2.1κ^{-1}. Hence, the effective surface tension of liquid marbles can be calculated using Equation (4.34) (Aussillous and Quéré 2006, Newton *et al.* 2007, Bormashenko *et al.* 2009a).

$$\gamma_{\text{eff}} = \frac{\rho g h^2}{4} \tag{4.34}$$

It is important to note that there are two independent sources of errors inherent in the puddle height method: (1) the maximal puddle height corresponding to the infinite volume of a liquid marble is unattainable, and (2) the actual apparent contact angles are significantly less than 180°. Both of them underestimate the value of the effective surface tension. However, the apparent contact angle affects it more than the height. It should be noted that accurate measurement of the apparent contact angle and the marble height is a bit challenging as it is difficult to determine the exact position of the liquid-particle interface on the marble. Keeping the errors associated with this method in mind, Bormashenko *et al.* (2009a) have calculated the effective surface tension of aqueous liquid marbles for values of θ_A between 130° and 140° using Equation (4.35), a modified form of (4.34). The values of γ_{eff} obtained were 13%–22% larger than those obtained using Equation (4.34).

$$\gamma_{\text{eff}} = \frac{\rho g h^2}{4 \sin^2 (\theta_A/2)} \tag{4.35}$$

The puddle height method has been used to determine the γ_{eff} of water and glycerol liquid marbles stabilized by fluorodecyltrichlorosilane-treated lycopodium (size ~30 μm) powdered particles and dichlorodimethylsilane-treated silica (size ~10 nm) powdered particles (Aussillous and Quéré 2004, 2006). For the former particles, they reported γ_{eff} values of 51 mN m^{-1} for water and 45 mN m^{-1} for glycerol while for the latter particles the values remained close to those of the encapsulated liquids. The γ_{eff} of the water liquid marbles, stabilized by the treated lycopodium particles, was further measured by Newton *et al.* (2007) and Bormashenko *et al.* (2009a) using the asymptotic height method. They, independently, reported values of γ_{eff} as 53 ± 5 mN m^{-1} and 50 ± 5 mN m^{-1}, respectively. Different particles affect the value of γ_{eff} of a given liquid marble differently depending on the types of interactions between the particles once on the liquid drop surfaces. This is illustrated in Table 4.2 for water liquid marbles (at ordinary temperature) stabilized by PTFE (size 100–200 nm) powder,

TABLE 4.2
Values of the Effective Surface Tension γ_{eff} of Water Liquid Marbles (at Ordinary Temperature), Stabilized by Powdered Particles of Different Surface Energy (Given in Terms of $\gamma_{sa}{}^d$ and $\gamma_{sa}{}^p$), Measured Using Different Methods

| | Particle Surface Energy Component/mN m^{-1} | | γ_{eff}/mN m^{-1} (From Different Methods) | | |
| | | | Maximum | Marble Shape | |
Marbles	$\gamma_{sa}{}^d$	$\gamma_{sa}{}^p$	Height	Analysis	Vibration
PVDF	28.1[a]	4.8[a]	70 ± 7	79 ± 5	75 ± 3
PE	30.0[b]	1.3[b]	66 ± 5	63 ± 3	60 ± 4
PTFE	17.0[b]	0.6[b]	60 ± 6	53 ± 5	53 ± 3
Lycopodium	–	–	50 ± 5	62 ± 5	43 ± 3

Source: Bormashenko, E. *et al., Colloids Surf. A* 351, 78–82, 2009.
[a] Values from Carre, A., *J. Adhesion Sci. Technol.*, 21, 961, 2007.
[b] Values from Clint, J.H. and Wicks, A.C., *Int. J. Adhes. Adhes.*, 21, 267–273, 2001.

polyethylene (PE) powdered particles (diameter ~2 μm), and PVDF nanobeads of average diameter 130 nm (Bormashenko *et al.* 2009a). The values of the γ_{eff} were obtained from the puddle height, marble shape analysis and marble vibration methods (Bormashenko *et al.* 2009a). Upon adsorption of the particles on the water drop surfaces, part of the particles is exposed to an aqueous environment leading to ionization of surface charges while the other part remains largely in the air phase and is not ionized. The degree of particle ionization depends on the polar nature of the particle, quantified by the magnitude of the polar component $\gamma_{sa}{}^p$ of the particle surface energy. For the PVDF, PE, and PTFE particles, the polar and dispersion components $\gamma_{sa}{}^d$ of their surface energy are given in Table 4.2. The ionized part of the particles plays a key role in the interaction of the particles and for the PVDF, PE, and PTFE particles, the degree of ionization decreases as the $\gamma_{sa}{}^p$ decreases. Consequently, the degree of repulsive interaction between the particles will decrease, and hence, γ_{int} becomes more negative. This accounts for the reason why the effective surface tension of the liquid marbles decreased as the polar component of the particle surface energy decreases.

Example 4.5: Calculating the Effective Surface Tension of Liquid Marbles Using the Puddle Height Method

The apparent contact angle of a PVDF particle-stabilized water liquid marble (10 μL) resting on a substrate at ambient conditions is 150°. If the surface tension of the encapsulated water is 72 mN m^{-1} and its density is 0.997 g cm^{-3} and the gravitational acceleration is 9.8 m s^{-2}, calculate the effective surface tension of the liquid marble.

Method

First, combine Equations (4.33) and (4.34) or (4.33) and (4.35). Second, substitute the necessary data (given) into the resultant equations and calculate the effective surface tension at once.

Answer

Combining Equations (4.33) and (4.34) gives

$$\gamma_{eff} = \rho g (\kappa^{-1})^2 \sin^2\left(\frac{\theta_A}{2}\right) = \rho g \times \frac{\gamma_{la}}{\rho g} \sin^2\left(\frac{\theta_A}{2}\right) = \gamma_{la} \sin^2\left(\frac{\theta_A}{2}\right)$$

$$= 72 \text{ mN m}^{-1} \times \sin^2\left(\frac{150}{2}\right) = 67 \text{ mN m}^{-1}$$

Combining Equations (4.33) and (4.35) gives

$$\gamma_{eff} = \rho g (\kappa^{-1})^2 = \rho g \times \frac{\gamma_{la}}{\rho g} = \gamma_{la} = 72 \text{ mN m}^{-1}$$

4.3.3.2 The Analysis of Marble Shape

The effective surface tension of liquid marbles can be obtained from the analysis of their shape (Aussillous and Quéré 2006, Bormashenko *et al.* 2009a,b). The precise shape of the marble is calculated only numerically in relation to an oblate spheroid since the shape of a liquid marble deformed by gravity is described satisfactorily by the shape of an oblate spheroid (Whyman and Bormashenko 2009). The fitting of the calculated and measured geometrical parameters allow the establishment of the effective surface tension of the liquid marble (Bhosale *et al.* 2008, Bormashenko *et al.* 2009a). This method is, however, complicated.

4.3.3.3 The Vibration of Liquid Marbles

The resonance frequencies of vibrated liquid marbles also provide a measure of their effective surface tension (Bormashenko 2011). Equation (4.36) is used to calculate γ_{eff} from this technique.

$$\gamma_{eff} = \frac{\rho V_m f_m^2}{2\pi \, h(\theta_A) \, (1 - \cos\theta_A)} \qquad (4.36)$$

The f_m is the resonance frequency of the marble, V_m is the volume of the liquid marble and $h(\theta_A)$ is a numerical multiplier whose values depend on the apparent contact angle of the marble as shown in Figure 4.8 (Celestini and Kofman 2006). (Other symbols retain their known meaning.) As can be seen, the value of $h(\theta_A)$ is close to 1 for $\theta_A \sim 90°$ and decreases as θ_A increases and eventually becomes zero at $\theta_A \sim 180°$. In real situations, θ_A does not reach 180°. It is worth mentioning that vibration experiments are impossible in cases where the deposited liquid marble rolls off the surface upon vibration (Bormashenko *et al.* 2009a). This usually occurs when the surface does not pin the contact area of the marble (Bormashenko *et al.* 2009a).

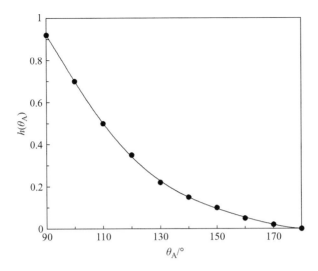

FIGURE 4.8 Possible numerical values of $h(\theta_A)$ for contact angles ranging from 90° to 180°. (From Celestini, F. and Richard, K., *Phys. Rev. E*, 73, 041602, 2006.)

Example 4.6: Calculating the Resonance Frequency of a Marble from the Effective Surface Tension

Taking the effective surface tension of the liquid marble in Example 4.5 as 67 mN m^{-1}, calculate the resonance frequency of the marble if $h(150°)$ is 0.1.

Method

Make the resonance frequency f_m the subject of the formula in Equation (4.36) and substitute $\rho = 997$ kg m^{-3}, $V_m = 1.0 \times 10^{-8}$ m^3, $h(150°) = 0.1$, $\gamma_{eff} = 67 \times 10^{-3}$ N m^{-1} and $\theta_A = 150°$ into the resultant formula and calculate it at once.

Answer

$$f_m = \sqrt{\frac{\gamma_{eff} \times 2\pi\, h(\theta_A)\, (1-\cos\theta_A)}{\rho V_m}} = \sqrt{\frac{67\times10^{-3}\ \text{N m}^{-1}\times2\pi\times0.1\times(1-\cos150°)}{997\ \text{kg m}^{-3}\times1.0\times10^{-8}\ \text{m}^3}}$$

$$= 89\ \text{s}^{-1}$$

4.3.3.4 The Wilhelmy Plate and Capillary Rise Methods

The Wilhelmy plate is also used to measure the effective surface tension of liquid marbles. The capillary rise and Wilhelmy plate methods have been used to measure the effective surface tension of water liquid marbles stabilized by PTFE powders of different particle sizes (1, 35 and 100 µm) (Arbatan and Shen 2011). With the capillary rise technique, a glass capillary tube is inserted into a marble and the height h_w to which the encapsulated water rise in the capillary tube is measured. The Laplace pressure exerted by the water liquid marble is directly measured by comparing the heights of the capillary rise from the marble and from a flat water surface in a beaker.

Equation (4.37), which is based on the Marmur's model, is then used to calculate the effective surface tension of the marble.

$$\gamma_{\text{eff}} = \frac{\rho g \Delta h_{\text{w}}}{1/R'} = \rho g \Delta h_{\text{w}} R' \qquad (4.37)$$

The Δh_{w} is the difference between the capillary rise from the flat water surface and that from the marble, ρ is the density of water, g is the acceleration due to gravity and $1/R'$ is the hydraulic radius defined as

$$\frac{1}{R'} = \frac{1}{R_1} + \frac{1}{R_2} \qquad (4.38)$$

The R_1 and R_2 are the principal radii of curvature of the liquid marble. This method does not require the apparent contact angle of the marble with the substrate. It is a simple and efficient method for determining the effective surface tension of liquid marbles.

For the Wilhelmy plate method, the surface tension of a flat water surface covered with powdered PTFE particles is calculated by using Equation (4.39) after measuring the wetting force F.

$$\gamma_{\text{eff}} = \frac{F}{2(L+t) \times \cos\theta} \qquad (4.39)$$

The L is the width of the plate, t is the thickness of the plate, and θ (normally 0°) is the contact angle between water and the plate. This method is illustrated in Figure 4.9 for two different PTFE particle sizes. The fundamental assumption here is that the

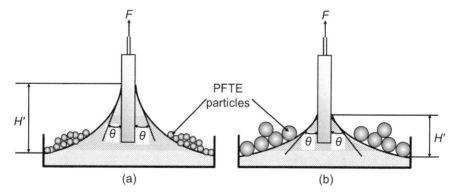

FIGURE 4.9 Schematic illustration of the Wilhelmy plate method of measuring the effective surface tension of a liquid surface, *e.g.* water, covered with hydrophobic particles like PFTE of relatively (a) small and (b) large size and the effect of particle size on the meniscus height H'. (From Arbatan, T. and Shen, W. *Langmuir*, 27, 12923–12929, 2011.)

TABLE 4.3

The Effective Surface Tension γ_{eff} (in mN m^{-1}) of Water Liquid Marbles (Different Volumes at Ordinary Temperature) Determined by the Capillary Rise and the Wilhelmy Plate Methods. The Surface Tension of Water (At Ordinary Temperature) Measured by the Capillary Rise and the Wilhelmy Plate Methods is ~71 and 72.3 mN m^{-1}, Respectively

Size of PTFE Particles/μm	γ_{eff} from the Capillary Rise (different volumes of marbles, given in μL)					γ_{eff} from Wilhelmy Plate
	30	50	100	200	300	
1	71	71	70	69	71	70.2
35	70	70	70	68	70	69.9
100	69	68	69	69	70	57.6

Source: Arbatan, T. and Shen, W., *Langmuir* 27, 12923–12929, 2011.

effective surface tension of a flat water surface covered by a layer of inert hydrophobic powder is the same as that of an aqueous marble covered by the same powder. This method offers a new way of investigating the properties of a liquid marble shell.

Values of the effective surface tension of the liquid marbles, obtained from the capillary rise and the Wilhelmy plate methods are given in Table 4.3. For the capillary rise method, the effective surface tension is generally independent of the volume of the liquid marbles and size of the PTFE particles encapsulating them. For the Wilhelmy plate method, however, the effective surface tension decreases as the size of the PTFE particles increases. This is interpreted in terms of the meniscus height. The relatively small PTFE particles are thought to be pushed away from the meniscus due to their size compared to the relatively large ones, leading to a higher meniscus height for the small particles and a lower meniscus height for the large particles. Measurements show that the wetting force was higher for the relatively small particles compared to the large ones.

4.3.3.5   The Pendant Marble Method

This method is based on the pendant drop method of measuring the surface tension of liquids described vividly in Chapter 1 (Section 1.3.4.1). The effective surface tension of water liquid marbles (at ordinary temperature) stabilized by different solid particles have been obtained using this method (Bormashenko *et al.* 2013). To obtain a pendant liquid marble, a syringe was filled with water and fixed vertically. The required volume of water was then suspended on the syringe's needle. A solid substrate covered with a thin layer of the powdered particles was carefully brought in contact with the suspended water drop and moved gently relative to it. The particles encapsulated the water drop to form a suspended marble (Bormashenko *et al.* 2013). The γ_{eff} of the marble was then measured with the Rame-Hart goniometer (Model 500) using the pendant drop method. The pendant liquid marble was imaged and the γ_{eff} was considered as a fitting parameter. The value of the γ_{eff} was adjusted until the solution of Equation (4.40) agrees with the experimental result obtained with the marble imaging (Bormashenko *et al.* 2013).

$$\gamma_{eff}\left(\frac{1}{R_1} + \frac{\sin\phi}{X}\right) = \frac{2\gamma_{eff}}{b} - \Delta\rho g z \qquad (4.40)$$

The values of the γ_{eff} measured depended strongly on the volume of the liquid marble and exhibited significant hysteresis (*i.e.* values were different under inflation and deflation of the marbles). This resulted from the degree of particle coverage (Bormashenko *et al.* 2013). Inflation of the marble increases its volume as well as the surface area and decreases the degree of particle coating while deflation of the marble does the opposite. The former decreases particle interactions while the latter increases it and thus leading to different γ_{eff} values as discussed previously. Upon inflation, values as high as 72 mN m^{-1} were obtained.

4.3.4 ELECTROWETTING AND MAGNETOWETTING OF LIQUID MARBLES

The wettability and shape of a liquid drop changes in the presence of an applied electric field and this phenomenon is known as electrowetting. During the process, the wettability of the liquid drop increases while the interfacial energy decreases (Vallet *et al.* 1996). Just like with liquid drops, an electric field can deform, move, or change the wettability of a liquid marble. The electrowetting of liquid marbles has been reported by Aussillous and Quéré (2006), Newton *et al.* (2007), Bormashenko *et al.* (2011), and many others. In the first, aqueous liquid marbles containing iron and coated with lycopodium or coated with iron and lycopodium particles were approached with a charged Teflon stick. The liquid marbles were observed to move with successive bouncing. The marbles were seen to eject tiny liquid drops during their motion. In the second, it was shown, using 0.01 M aqueous KCl solution liquid marbles stabilized with hydrophobic lycopodium particles, that the electrowetting of liquid marbles increases their contact area with the substrate and decreases the apparent contact angle. However, provided the applied voltage is not too large, these changes are completely reversible as the marbles return to their initial contact area and initial apparent contact angle when the electric current is removed. Finally, large voltages were reported to burst the marbles. The deformation of aqueous liquid marbles, stabilized by PVDF, at a threshold voltage is reported in the third. Marble deformation was impossible below the threshold voltage. Apart from electrowetting, two contacting liquid marbles of aqueous NaCl solution have been forced to coalesce using electrical charging (Liu *et al.* 2017). A DC electric field was used for the process. The threshold voltage leading to coalescence was seen to depend on the stabilizing particles and the surface tension of the aqueous phase. For multiple marbles, coalescence was driven by a sufficiently high threshold voltage that increases linearly with the number of marbles.

The use of a magnetic field to change the wettability of a liquid marble as well as its shape is known as magnetowetting. In order to respond to the magnetic field, the liquid marble is expected to have a magnetic coating or content like magnetite. Liquid marbles with magnetic content are known as ferrofluid marbles. For liquid marbles coated with magnetic particles, the force F'_{mag} acting on each individual magnetic particle is given by Equation (4.41) (Zhao *et al.* 2012).

$$F'_{mag} = \frac{V_p \Delta \chi}{\mu_o} (\nabla B) B \qquad (4.41)$$

The V_p is the volume (m³) per powdered particle, $\Delta \chi$ is the difference between the magnetic susceptibility of the powdered particle and that of the surrounding medium, μ_o is the permeability of vacuum and has a value of $4\pi \times 10^{-7}$ T m A^{-1}, B is the magnetic flux density (T), and ∇B is the magnetic field gradient (T m^{-1}). From Equation (4.41), the overall magnetic force F_{mag} acting on a liquid marble is

$$F_{mag} = \frac{m' \Delta \chi}{\rho' \mu_o} (\nabla B) B \qquad (4.42)$$

in which m' is the total mass of magnetic powdered particles on the marble and ρ' is the density of the particles.

Example 4.7 Estimating the Magnitude of the Magnetic Force Acting on a Magnetic Liquid Marble

A magnetic aqueous liquid marble (15 μL) is prepared by wrapping the liquid drop with Fe_3O_4 nanoparticles at ambient conditions. Given that the magnetic susceptibility of the surrounding medium is negligibly small and the density of an Fe_3O_4 nanoparticle, its magnetic susceptibility and the total mass of magnetic particles on the marble are 5.18×10^3 kg m^{-3}, 1.45 and 4.9×10^{-7} kg, respectively, estimate the magnitude of the magnetic force acting on the liquid marble for a magnetic flux density of 15.2 mT and magnetic field density of 1.76 T m^{-1}.

Method

Substitute the data given into Equation (4.42) and solve for the magnitude of the magnetic force F_{mag} acting on the marble.

Answer

$$F_{mag} = \frac{m'\Delta\chi}{\rho'\mu_o}(\nabla B)B = \frac{4.9\times10^{-7}\ kg\times1.45\times1.76\ T\ m^{-1}\times15.2\times10^{-3}\ T}{5.18\times10^3\ kg\ m^{-3}\times4\pi\times10^{-7}\ T\ m\ A^{-1}}$$

$$= 2.92\times10^{-6}\ \overbrace{N\ A^{-1}\ m^{-1}}^{I}\times\ A\ m = 2.92\ \mu N$$

Although the magnitude of the force is very small, it is reported (Zhao *et al.* 2012) to actuate the liquid marble provided the magnate is 16 mm from the marble.

Ferrofluid liquid marbles are known to deform under the influence of a permanent magnet. The magnetic Bond number Bm is used to characterize the marble height as well as the contact radius in terms of the dimensionless height $h^* = h/R_m$ and the dimensionless contact radius $\delta^*_m = \delta_m/R_m$ in which R_m is the radius of the marble prior to deformation. For the case where the magnetic force is much larger than the weight of the marble and the magnetic flux density imparted onto the marble is much larger than the critical magnetic flux density, it is shown (Nguyen 2013) that $\delta^*_m \sim Bm^{1/4}$ and $h^* \approx 2$ for small quasi-spherical marbles while $\delta^*_m \sim Bm^{1/4}$ and $h^* \approx Bm^{-1/2}$ for large puddle-shaped ones. The $Bm = R_m M_F B/\gamma_{eff}$ in which M_F is the magnetization density of the ferrofluid.

4.3.5 EVAPORATION AND FREEZING OF LIQUID MARBLES

The presence of particles on the surfaces of liquid marbles affects the evaporation of the encapsulated liquid drop and the behavior of the liquid drop upon freezing. The particle coating on a liquid marble reduces the rate of diffusion-controlled evaporation of the encapsulated liquid compared to an uncoated drop. This depends on the degree of particle coating: the higher the degree of particle coating, the lower the rate of evaporation of the encapsulated liquid and vice versa. Generally, the rate of evaporation is relatively high for relatively high vapor pressure (or highly volatile) liquids compared with relatively low vapor pressure (less volatile) ones (Binks and Tyowua 2013). Liquid marbles buckle and crumple eventually as the encapsulated liquid evaporates because the particles encapsulating the liquid drop progressively get forced together (Dandan and Erbil 2009, Tosun and Erbil 2009, Doganci *et al.* 2011) as shown in Figure 4.10. Hollow spherical shells are sometimes left after complete evaporation of the encapsulated liquid (Hapgood and Khanmohammadi 2009, Eshtiaghi *et al.* 2010). The evaporation of water liquid marbles stabilized by PTFE particles of size 7–12 μm and hydrophobic fumed silica particles in the form of aggregates (200–500 nm) of long chain (3.27 nm) beads treated with either hexamethyldisilazane (HMDS) or dichlorodimethylsilane have been studied (Bormashenko *et al.* 2010). The aggregated chains of fumed silica particles allowed the formation of particulate network on the liquid marble surfaces. This increased the robustness of the marbles against rupture under an applied force and also delayed buckling upon evaporation compared to the PTFE-stabilized ones.

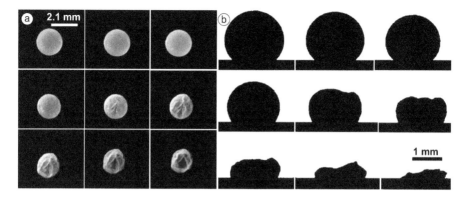

FIGURE 4.10 Photographs of (a) top and (b) side view of a water liquid marble (5 μL) stabilized by PTFE particles during evaporation at 21–25°C and relative humidity of 54%, showing the eventual buckling and crumpling of the liquid marble as the encapsulated water evaporates. (From Tosun, A. and Erbil, H.Y., *Appl. Surf. Sci.*, 256, 1278–1283, 2009.)

The response of a liquid marble to freezing is different from that of a bare liquid drop as mentioned previously. For water liquid marbles, shape transformation occurs upon freezing (Hashmi *et al.* 2012, Zang *et al.* 2014). Water liquid marbles stabilized by quasi-spherical hydrophobic lycopodium particles (average diameter ~28 μm) were placed on a silicon wafer maintained at −8°C (Hashmi *et al.* 2012). The shape of the liquid marble changed from quasi-spherical to a bell-shaped and then finally to a flying saucer-shaped morphology as the freezing front moved gradually from the silicon wafer towards the top of the liquid marble (Hashmi *et al.* 2012). This is in contrast to what is seen for a bare water drop. The observation was attributed to the preferred nucleation induced by the lycopodium particles on the marble and Marangoni convection (Hashmi *et al.* 2012). Recall that Marangoni convection (also thermo-capillary convection) is mass transfer along an interface due to temperature gradient; *cf.* Gibbs-Marangoni effect, which is due to surface tension gradient. Strictly speaking, temperature gradient leads to surface tension gradient. Using fumed silica particles (diameter ~20 nm) of varying degrees of hydrophobicity (quantified by the % of surface SiOH group), the dependence of water marble shape on particle hydrophobicity upon freezing was witnessed (Zang *et al.* 2014). The % of SiOH group on the particle surfaces was in the range of 14% (most hydrophobic) to 61% (least hydrophobic). The freezing experiments were performed at a relative humidity of ~40% and temperature of ~ −15°C on 15 and 20 μL liquid marbles and compared to that on bare water drops on a superhydrophobic surface. For liquid marbles stabilized by the most hydrophobic particles; a vertically prolonged morphology with a pointed protrusion on the top was observed on freezing, similar to a bare water drop on a superhydrophobic surface as shown in Figure 4.11. The vertical prolonged effect diminishes as particle hydrophobicity decreases. For marbles stabilized by the less hydrophobic particles, however, a change from quasi-spherical to a lateral expanded flying

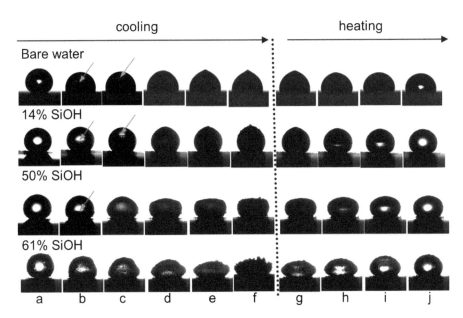

FIGURE 4.11 Shape transformation of a bare water drop and water liquid marbles (15 μL) upon freezing (a–f) and re-melting (g–j). The marbles are stabilized by fumed silica particles of different degrees of hydrophobicity (given in terms of % SiOH). The arrows indicate the advancing ice-water interface upwards. The time scale involved in freezing and re-melting is ~90 s. (From Zang, D. *et al.*, *Soft Matter*, 10, 1309–1314, 2014.)

saucer-shaped morphology was observed (Figure 4.11). This indicates that the particle coating on the surfaces of liquid marbles may drastically alter their response to the environment. The different responses to freezing were interpreted in terms of the different heterogeneous nucleation sites owing to the different positions of the particles at the water-air interface as shown in Figure 4.12. If the particles project more in the water phase (less hydrophobic particles), the formation of ice embryos tend to occur in the concave cavities between the particles (Figure 4.12). The volume expansion of water as a result of freezing and continuous nucleation lead to continuous lateral stretching of the particle network encapsulating the liquid drop surface and ultimately to the horizontally inflated shape of the marble. Nonetheless, if the particles project more in the air phase (most hydrophobic particles), nucleation takes place on the convex surface of the particles (Figure 4.12), just like that of a bare water drop on a hydrophobic substrate (Zang *et al.* 2014). In both cases, the quasi-spherical shape of the marbles was recovered on re-melting, with a relatively small water leakage for those stabilized by the least hydrophobic particles (50% and 61% SiOH). The effect of temperature gradient on the marbles was investigated by placing them in a freezer maintained at about −15°C where they would be at a more uniform external temperature field and thereby reducing the temperature gradient in the marble (Zang *et al.* 2014). This yielded the same

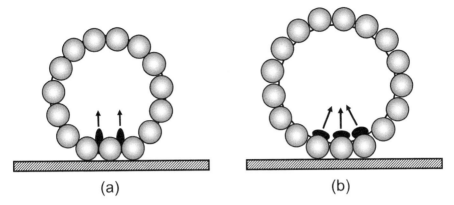

(a)   (b)

FIGURE 4.12   Different nucleation sites in water liquid marbles stabilized by fumed silica particles of different hydrophobicities. (a) For marbles stabilized by the least hydrophobic particles (61% SiOH), nuclei formation is in the concave cavities between the silica particles. (b) For the liquid marbles stabilized by the very hydrophobic particles (14% SiOH), ice nucleation (black) begins on the convex silica particle surfaces. (From Zang, D. *et al., Soft Matter,* 10, 1309–1314, 2014.)

shape transformation as discussed previously. This indicates that temperature gradient and hence Marangoni convection plays little or no role in the shape transformation observed in the marbles.

Lastly, it was shown (Zang *et al.* 2014) that the Gibbs free energy barrier and the site of heterogeneous nucleation depend on particle wettability and play an important role on the shape transformation of the liquid marbles on freezing. Relatively hydrophilic particles significantly reduce the energy barrier for nucleation, giving rise to ice nucleation between the particles cavities. The energy barrier is, however, high for very hydrophobic particles and leads to ice nucleation on the particle surfaces.

4.4   CONCLUSION

This chapter first presents the principles behind liquid marble formation and then describes how liquid marbles are prepared, naming the various powdered particles that have been used. This is followed by a description of the dynamic and static properties of liquid marbles, which the uses of liquid marbles is based on. In doing this, particular attention is given to the effective surface tension of liquid marbles and its origin and measurement are thoroughly described. The chapter ends with the behavior of liquid marbles upon evaporation and freezing. It must be understood that the applications of liquid marbles discussed in Chapter 5 is based on this chapter and as such, the reader is encouraged to understand it before moving to Chapter 5.

EXERCISES

DISCUSSION QUESTIONS

Question 1
a. What are the underlying principles of liquid marble formation?
b. Describe how liquid marbles are prepared.
c. Differentiate between Bond, Weber, and Ohnesorge numbers. What is the significance of these numbers in liquid marble technology?

Question 2
a. Differentiate between the 'preformed droplet template' and 'mechanical dispersion' methods of liquid marble formation.
b. Describe at least ten particles used in the preparation of liquid marbles.

Question 3
a. Describe the statics and dynamics of liquid marbles.
b. Describe the behavior of aqueous liquid marbles in the presence of an electric field.

Question 4
a. Describe the various ways through which the effective surface tension of liquid marbles can be measured.
b. Explain how liquid marbles evaporate and respond to freezing.

NUMERICAL QUESTIONS

Question 1
a. Calculate the length of particle protrusion into the air and liquid drop (50 µL) phases given that the particle (radius 2 µm) makes a three-phase contact angle of 140° with its surfaces upon adsorption. Calculate the diameter of the drop given that it is quasi-spherical.
b. Show that the Gibb's free energy of a powdered particle (radius R_p) adsorbed at a planar liquid-air interface of interfacial tension γ_{la} is

$$\Delta G = \pi R_p^2 \gamma_{la} (1 - |\cos\theta|)^2$$

in which θ is the three-phase contact angle.

Question 2

a. Calculate the effective surface tension of an aqueous liquid marble given that the apparent contact angle when resting on a substrate is 140°. Take the surface tension and density of the encapsulated water and the acceleration of gravity as 72 mN m^{-1}, 0.997 g cm^{-3} and 9.8 m s^{-2}, respectively.

b. In the Wilhelmy plate method of effective surface tension measurement using a coverslip of length 18 mm and thickness 0.1 mm, the wetting force was reported to be 2.63 mN for a water surface covered with PTFE particles of size 1 μm when the contact angle is 0°. Calculate the effective surface tension of the water surface.

Question 3

a. Show that the condition on the viscosity η and the inclination angle α in the Mahadevan-Pomeau regime, where the rotation of small spherical drops is considered solid-like owing to a small Reynolds number is

$$\eta^2 > \rho \gamma_{la} \kappa^{-1} \alpha$$

in which all symbols retain their known meaning.

b. Given that

$$We = \frac{\rho V'^2 R_m}{\gamma_{la}} < 1 \text{ and } Ca = \frac{\eta V'}{\gamma_{la}} < 1$$

where all symbols maintain their known meaning, show that

$$R_m > \alpha \kappa^{-1}$$

FURTHER READING

Bormashenko, E.Y. *Wetting of Real Surfaces*. Berlin, Germany: Walter de Gruyter GmbH, 2013.
Ooi, C.H. and N.-T. Nguyen. "Manipulation of Liquid Marbles." *Microfluid. Nanofluid.* 19 (2015): 483–95.

REFERENCES

Agland, S. and S.M. Iveson. "The Impact of Liquid Drops on Powder Bed Surfaces." *CHEMECA* 99 (1999): 218–24.
Arbatan, T. and W. Shen. "Measurement of the Surface Tension of Liquid Marbles." *Langmuir* 27 (21) (2011): 12923–29.
Aussillous, P. and D. Quéré. "Liquid Marbles." *Nature* 411 (2001): 924–27.
Aussillous, P. and D. Quéré. "Properties of Liquid Marbles." *Proc. R. Soc. Chem. Lond. A* 462 (2006): 973–99.
Aussillous, P. and D. Quéré. "Shapes of Rolling Liquid Drops." *J. Fluid Mech.* 512 (2004): 133–51.
Bhosale, P.S. and M.V. Panchagnula. "On Synthesizing Solid Polyelectrolyte Microspheres from Evaporating Liquid Marbles." *Langmuir* 26 (13) (2010): 10745–49.
Bhosale, P.S., M.V. Panchagnula and H.A. Stretz. "Mechanically Robust Nanoparticle Stabilized Transparent Liquid Marbles." *Appl. Phys. Lett.* 93 (3) (2008): 034109-09-3.
Binks, B.P. and A.T. Tyowua. "Influence of the Degree of Fluorination on the Behavior of Silica Particles at Air-Oil Surfaces." *Soft Matter* 9 (3) (2013): 834–45.
Binks, B.P. and R. Murakami. "Phase Inversion of Particle-Stabilized Materials from Foams to Dry Water." *Nat. Mater.* 5 (2006): 865–69.
Binks, B.P., A.N. Boa, M.A. Kibble, G. Mackenzie and A. Rocher. "Sporopollenin Capsules at Fluid Interfaces: Particle-Stabilized Emulsions and Liquid Marbles." *Soft Matter* 7 (8) (2011): 4017–24.
Binks, B.P., S.K. Johnston, T. Sekine and A.T. Tyowua. "Particles at Oil-Air Surfaces: Powdered Oil, Liquid Oil Marbles, and Oil Foam." *ACS Appl. Mater. Interfaces* 7 (26) (2015): 14328–37.
Binks, B.P., T. Sekine and A.T. Tyowua. "Dry Oil Powders and Oil Foams Stabilized by Fluorinated Clay Platelet Particles." *Soft Matter* 10 (4) (2014): 578–89.
Bormashenko, E. "Liquid Marbles: Properties and Applications." *Curr. Opin. Colloid Interface Sci.* 16 (4) (2011): 266–71.
Bormashenko, E., A. Musin, G. Whyman *et al.* "Revisiting the Surface Tension of Liquid Marbles: Measurement of the Effective Surface Tension of Liquid Marbles with the Pendant Marble Method." *Colloids Surf. A* 425 (2013b): 15–23.
Bormashenko, E., R. Balter and D. Aurbach. "Micropump Based on Liquid Marbles." *Appl. Phys. Lett.* 97 (9) (2010b): 091908-08-2.
Bormashenko, E., R. Pogreb, A. Musin, R. Balter, G. Whyman and D. Aurbach. "Interfacial and Conductive Properties of Liquid Marbles Coated with Carbon Black." *Powder Technol.* 203 (3) (2010a): 529–33.
Bormashenko, E., R. Pogreb, G. Whyman and A. Musin. "Jetting Liquid Marbles: Study of the Taylor Instability in Immersed Marbles." *Colloid Polym. Sci.* 291 (6) (2013a): 1535–39.
Bormashenko, E., R. Pogreb, G. Whyman and A. Musin. "Surface Tension of Liquid Marbles." *Colloids Surf. A* 351 (2009a): 78–82.
Bormashenko, E., R. Pogreb, G. Whyman, A. Musin, Y. Bormashenko and Z. Barkay. "Shape, Vibrations, and Effective Surface Tension of Water Marbles." *Langmuir* 25 (4) (2009b): 1893–96.
Bormashenko, E., Y. Bormashenko, A. Musin and Z. Barkay. "On the Mechanism of Floating and Sliding of Liquid Marbles." *Chem. Phys. Chem.*, 10 (2009d), 654–656.
Bormashenko, E., R. Pogreb, T. Stein, G. Whyman, M. Schiffer and D. Aurbach. "Electrically Deformable Liquid Marbles." *J. Adhes. Sci. Technol* 25 (12) (2011): 1371–77.

Bormashenko, E., R. Pogreb, Y. Bormashenko, A. Musin and T. Stein. "New Investigations on Ferrofluidics: Ferrofluidic Marbles and Magnetic-Field-Driven Drops on Superhydrophobic Surfaces." *Langmuir* 24 (21) (2008): 12119–22.

Bormashenko, E., Y. Bormashenko and A. Musin. "Water Rolling and Floating Upon Water: Marbles Supported by Water/Marble Interface." *J. Colloid Interface Sci.* 333 (2009c): 419–21.

Bormashenko, E., Y. Bormashenko and G. Oleg. "On the Nature of the Friction between Nonstick Droplets and Solid Substrates." *Langmuir* 26 (15) (2010b): 12479–82.

Brown, R.A. and L.E. Scriven. "The Shape and Stability of Rotating Liquid Drops." *Proc. R. Soc.* 371 (1746) (1980): 331–57.

Carre, A. "Polar Interactions at Liquid/Polymer Interfaces." *J. Adhesion Sci. Technol.* 21 (2007): 961.

Celestini, F. and R. Kofman. "Vibration of Submillimeter-Size Supported Droplets." *Phys. Rev. E* 73 (4) (2006): 041602.

Chandrasekhar, S. "The Stability of a Rotating Liquid Drop." *Proc. R. Soc.* 286 (1404) (1965): 1–26.

Clint, J.H. and A.C. Wicks. "Adhesion under Water: Surface Energy Considerations." *Int. J. Adhes. Adhes.* 21 (2001): 267–73.

Dandan, M. and H.Y. Erbil. "Evaporation Rate of Graphite Liquid Marbles: Comparison with Water Droplets." *Langmuir* 25 (14) (2009): 8362–67.

Doganci, M.D., B.U. Sesli, H.Y. Erbil, B.P. Binks and I.E. Salama. "Liquid Marbles Stabilized by Graphite Particles from Aqueous Surfactant Solutions." *Colloids Surf. A* 384 (1–3) (2011): 417–26.

Dupin, D., S.P. Armes and S. Fujii. "Stimulus-Responsive Liquid Marbles." *J. Am. Chem. Soc.* 131 (15) (2009): 5386–87.

Dupin, D., K.L. Thompson and S.P. Armes. "Preparation of Stimulus-Responsive Liquid Marbles Using a Polyacid-Stabilized Polystyrene Latex." *Soft Matter* 7 (15) (2011): 6797–800.

Eshtiaghi, N. and K.P. Hapgood. "A Quantitative Framework for the Formation of Liquid Marbles and Hollow Granules from Hydrophobic Powders." *Powder Technol.* 223 (2012): 65–76.

Eshtiaghi, N., J.S. Liu and K.P. Hapgood. "Fromation of Hollow Granules from Liquid Marbles: Small Scale Experiments." *Powder Technol.* 197 (2010): 184–95.

Eshtiaghi, N., J.S. Liu, W. Shen and K.P. Hapgood. "Liquid Marble Formation: Spreading Coefficients or Kinetic Energy?" *Powder Technol.* 196 (2009): 126–32.

Feng, L., S. Li, Y. Li *et al.* "Super-Hydrophobic Surfaces: From Natural to Artificial." *Adv. Mater.* 14 (24) (2002): 1857–60.

Forny, L., I. Pezron and K. Saleh. "Dry Water: From Physico-Chemical Aspects to Process Related Parameters." In *9th International Symposium on Agglomeration.* Sheffield University, Sheffield, UK, 2009.

Fujii, S., M. Suzaki, S.P. Armes *et al.* "Liquid Marbles Prepared from pH-Responsive Sterically Stabilized Latex Particles." *Langmuir* 27 (13) (2011): 8067–74.

Gao, L., T.J. McCarthy and X. Zhang. "Wetting and Superhydrophobicity." *Langmuir* 25 (24) (2009): 14100–104.

Hapgood, K.P. and B. Khanmohammadi. "Granulation of Hydrophobic Powders." *Powder Technol.* 189 (2009): 253–62.

Hashmi, A., A. Strauss and J. Xu. "Freezing of a Liquid Marble." *Langmuir* 28 (28) (2012): 10324–28.

Hill, R.J.A. and L. Eaves. "Nonaxisymmetric Shapes of a Magnetically Levitated and Spinning Water Droplet." *Phys. Rev. Lett.* 101 (23) (2008): 234501.

Kim, S-H., S. Lee and S-M. Yang. "Janus Microspheres for a Highly Flexible and Impregnable Water-Repelling Interface." *Angew. Chem. Int. Ed.* 49 (14) (2010): 2535–38.

Lee, C.P., A.V. Anilkumar, A.B. Hmelo and T.G. Wang. "Equilibrium of Liquid Drops under the Effects of Rotation and Acoustic Flattening: Results from USML-2 Experiments in Space." *J. Fluid Mech.* 354 (1998): 43–67.

Liu, Z., X. Fu, B.P. Binks and H.C. Shum. "Coalescence of Electrically Charged Liquid Marbles." *Soft Matter* 13 (2017): 119–24.

Mahadevan, L. and Y. Pomeau. "Rolling Droplets." *Phys. Fluids* 11 (1999): 2449.

Matsukuma, D., H. Watanabe, H. Yamaguchi and A. Takahara. "Preparation of Low-Surface-Energy Poly[2-(Perfluorooctyl)Ethyl Acrylate] Microparticles and Its Application to Liquid Marble Formation." *Langmuir* 27 (4) (2011): 1269–74.

McEleney, P., G.M. Walker, I.A. Larmour and S.E.J. Bell. "Liquid Marble Formation Using Hydrophobic Powders." *Chem. Eng. J.* 147 (2–3) (2009): 373–82.

McHale, G., S.J. Elliott, M.I. Newton, D.L. Herbertson and K. Esmer. "Levitation-Free Vibrated Droplets: Resonant Oscillations of Liquid Marbles." *Langmuir* 25 (1) (2008): 529–33.

McHale, G., N.J. Shirtcliffe, M.I. Newton, F.B. Pyatt and S.H. Doerr. "Self-Organization of Hydrophobic Soil and Granular Surfaces." *Appl. Phys. Lett* 90 (2007): 054110.

Newton, M.I., D.L. Herbertson, S.J. Elliott, N.J. Shirtcliffe and G. McHale. "Electrowetting of Liquid Marbles." *J. Phys. D: Appl. Phys.* 40 (1) (2007): 20.

Nguyen, N.-T. "Deformation of Ferrofluid Marbles in the Presence of a Permanent Magnet." *Langmuir* 29 (2013): 13982–89.

Nguyen, H.-T., K.P. Hapgood and W. Shen. "Observation of Liquid Marble Morphology Using Confocal Microscopy." *Chem. Eng. J.* 201 (2010): 396–405.

Pike, N., D. Richard, R. Foster and L. Mahadevan. "How Aphids Lose Their Marbles." *Proc. R. Soc. Lond. B* 269 (2002): 1211–1215.

Quéré, D. and P. Aussillous. "Non-Stick Droplets." *Chem. Eng. Technol.* 25 (2002): 925–28.

Rayleigh, L. "The Equilibrium of Revolving Liquid under Capillary Force." *Phil. Mag.* 28 (164) (1914): 161–70.

Richard, D. and D. Quéré. "Viscous Drops Rolling on a Tilted Non-Wettable Solid." *Europhys. Lett.* 48 (3) (1999): 286.

Sivan, V., S.-Y. Tang, A.P. O'Mullane *et al.* "Liquid Metal Marbles." *Adv. Funct. Mater.* 23 (2) (2013): 144–52.

Tosun, A. and H.Y. Erbil. "Evaporation Rate of PTFE Liquid Marbles." *Appl. Surf. Sci.* 256 (5) (2009): 1278–83.

Tyowua, A.T. "Solid Particles at Fluid Interfaces: Emulsions, Liquid Marbles, Dry Oil Powders and Oil Foams." PhD thesis of the University of Hull, UK, 2014.

Vadivelu, R.K., C.H. Ooi, R.-Q. Yao *et al.* "Generation of Three-Dimensional Multiple Spheroid Model of Olfactory Ensheating Cells Using Floating Liquid Marbles." *Sci. Rep.* 5 (2015): 15083.

Vallet, M., B. Berge and L. Vovelle. "Electrowetting of Water and Aqueous Solutions on Poly(ethylene terephthalate) Insulating Films." *Polymer* 37 (12) (1996): 2465–70.

Verplanck, N., Y. Coffinier, V. Thomy and R. Boukherroub. "Wettability Switching Techniques on Superhydrophobic Surfaces." *Nanoscale Res. Lett.* 2 (12) (2007): 577–96.

Whitby, C.P., X. Bian and R. Sedev. "Spontaneous Liquid Marble Formation on Packed Porous Beds." *Soft Matter* 8 (44) (2012): 11336–42.

Whyman, G. and E. Bormashenko. "Oblate Spheroid Model for Calculation of the Shape and Contact Angles of Heavy Droplets." *J. Colloid Interface Sci.* 331 (1) (2009): 174–77.

Xue, Y., H. Wang, Y. Zhao *et al.* "Magnetic Liquid Marbles: A 'Precise' Miniature Reactor." *Adv. Mater.* 22 (43) (2010): 4814–18.

Zang, D., D. Kejun, W.W. Lin *et al.* "Tunable Shape Transformation of Freezing Liquid Water Marbles." *Soft Matter* 10 (2014): 1309–14.

Zhao, Y., Z. Xu, M. Parhizkar, J. Fang, X. Wang and T. Lin. "Magnetic Liquid Marbles, Their Manipulation and Applications in Optical Probing." *Microfluid. Nanofluid.* 13 (2012): 555–64.

5 Applications of Liquid Marbles

5.1 APPLICATIONS OF LIQUID MARBLES IN MICROFLUIDICS

Microfluidics is the science and technology of manipulating fluids at the microscale level in networks of channels, having dimensions in the range of tens to hundreds of micrometres, for different applications. With the realization of liquid marbles, small amount (≤ 1 µL) of fluids can be manipulated outside microchannels. This has been illustrated in many ways. Firstly, liquid marbles have been used for the transportation of small volumes of liquids without leakage from one point to another (Aussillous and Quéré 2001) without requiring microchannels. Secondly, ferrofluidic liquid marbles have been used as ferrofluidic bearing (Bormashenko *et al.* 2008). Aqueous liquid marbles (5–70 µL) containing 2.5–25 g L^{-1} of γ-Fe$_2$O$_3$ particles, stabilized by PVDF particles (average diameter 130 nm), were used as the ferrofluidic liquid marbles. The marbles exhibited apparent contact angle values as high as 140° on superhydrophobic surfaces. The marbles were said to be easily activated by an external magnetic field. For example, a magnetic field of 0.5 T supplied a velocity of 25 ± 3 cm s^{-1} to a 20 µL liquid marble (Bormashenko *et al.* 2008). By encasing the ferrofluidic liquid marbles (along with pure water ones) between polyethylene and polypropylene films (Figure 5.1), a ferrofluidic microfluidic device which can be easily activated by an external magnetic field was prepared. The confined marbles, between the two polymer films, acted as stable drop-based bearing.

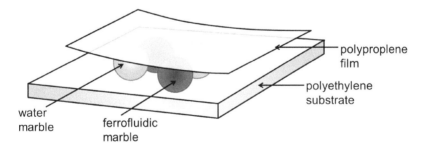

FIGURE 5.1 Schematic of a microfluidic device based on ferrofluidic marbles mixed with water ones. (Drawn from Bormashenko, E. *et al.*, *Langmuir*, 24, 12119–12122, 2008.)

5.2 APPLICATIONS OF LIQUID MARBLES IN THE ENVIRONMENTAL SCIENCE

Taking advantage of the permeability of the powder shell coating of liquid marbles, liquid marbles can be used for qualitative and quantitative gas sensing using an appropriate indicator, preferably one that changes color in the presence of the gas (Tian *et al.* 2010a). This has been demonstrated using phenolphthalein solution liquid marbles and aqueous liquid marbles containing $CuCl_2.2H_2O$ or $CoCl_2.2H_2O$, stabilized by PTFE powdered particles (Tian *et al.* 2010a, 2010b). The color of the liquid marbles changed rapidly upon exposure to NH_3 vapor from NH_4OH solution. The color of the phenolphthalein marble faded as it loses NH_3 after exposure, while that of the $CuCl_2.2H_2O$ and $CoCl_2.2H_2O$ marble did not due to stable complex formation with the transition metal ions. This shows that using an appropriate indicator, liquid marbles can be used as effective gas sensors. This will be of great importance in the environmental science as well as other related chemical industries. For example, liquid marbles have been used as effective gas sensors in the printing industry. Flexography, a dominant process in the printing of most packaging materials including corrugated boxes and polymer films, uses water-based ink formulations for polymer film printing. Inks formulated using the acrylic resin system are stable pigment suspension at moderately high pH (8.0–9.5) due to the dissociation of water in the presence of NH_3 and low molecular weight amines are used for flexographic printing. Upon printing, the drying of the ink is initiated by the evaporation of the NH_3 and the low molecular weight amines. During the drying process, the acrylic resin suspension system coagulates as the pH approaches values below 7, causing adhesion of the dried ink to the substrate. Unfortunately, the release of NH_3 and the low molecular weight amines are a major source of air pollution in the printing industry. Because Co^{2+} forms stable complexes with low molecular weight primary, secondary and some tertiary amines, a $CoCl_2$ solution liquid marble, stabilized by PTFE powdered particles, has been used to detect the presence of NH_3 and amine vapor in air, released from a flexographic ink (Tian *et al.* 2010a). Similarly, NH_4OH solution liquid marbles were exposed to dried circular paper discs treated with $CuCl_2$ solution to determine the amount of NH_3 vapor released (Tian *et al.* 2010a). In addition to qualitative and quantitative gas sensing, liquid marble can also be used to emit gas.

The ability of liquid marbles to float on liquid surfaces can also be made use of in environmental pollution monitoring. The floating ability of liquid marbles depends largely on the surface energy of the particles encasing them compared to that of the bulk liquid and the properties of the bulk liquid such as pH. Water liquid marbles coated with poly(2-vinylpyridine) have been shown to be pH-sensitive (Fujii *et al.* 2010). The marbles exhibited long-term stability when transferred onto the surface of liquid water (pH 4.9 or above). On the contrary, more acidic solutions (pH < 2.9) led to immediate disintegration (Fujii *et al.* 2010). This can be used as a qualitative check for the pH of aqueous solutions. In a related study, PVDF-stabilized water liquid marbles were shown to float stably on the surface of uncontaminated water but disintegrated on water surfaces contaminated with silicon oil and kerosene (Bormashenko and Musin 2009). This can be used in the remote sensing of water pollution caused by oil contamination in environmental monitoring.

5.3   LIQUID MARBLES AS MICROPUMPS

A pair of liquid marbles (of different volumes) stabilized with a given powdered particle can be connected with a capillary tube and used as a micropump (Bormashenko *et al.* 2010). Liquid marbles (of the same volume) stabilized with different particles can also be used for the experiment (Bormashenko *et al.* 2010). This idea is based on the difference in the Laplace pressure in the marbles owing to the difference in volume of the marbles or effective surface tension γ_{eff} of the marbles (if different particles are used). Due to the difference in pressure, materials flow from the relatively small liquid marble to the relatively big one or from a marble of higher γ_{eff} to the one of lower γ_{eff} when connected with a capillary tube. The flow of materials stops when the pressure in both marbles is approximately equal (Bormashenko *et al.* 2010). This observation emphasizes the importance of pressure gradient in the experiment. The Laplace pressure in a marble is $\sim 2\gamma_{eff}/R_m$, where R_m is the radius of the marble. Because areas of particle defect usually exist on the surfaces of liquid marbles, it is reasonable to say that the encapsulated drop interacts with the atmosphere. Thus the total pressure in the marble is $\sim(P_o + 2\gamma_{eff}/R_m)$, where P_o represents the atmospheric pressure. Therefore, the pressure difference ΔP_m between two liquid marbles of the same volume, stabilized by different particle types is given in Equation (5.1) (Bormashenko *et al.* 2010).

$$\Delta P_m \approx \frac{2\,|\,\Delta\gamma_{eff}\,|}{R_m} \tag{5.1}$$

In a case where the volumes of the liquid marbles are different, ΔP_m can be estimated using Equation (5.2)

$$\Delta P_m \approx \left[\left(\frac{2\gamma_{eff}}{R_{ms}}\right)_{small} - \left(\frac{2\gamma_{eff}}{R_{mL}}\right)_{large}\right] \tag{5.2}$$

provided they are stabilized by the same particle type. The $\Delta\gamma_{eff}$ is the difference in the effective surface tension of the liquid marbles. The $(2\gamma_{eff}/R_{ms})_{small}$ represents the pressure contribution of the small marble while $(2\gamma_{eff}/R_{mL})_{large}$ represents that from the large one. The R_{ms} and R_{mL} represent the radius of the small and large liquid marbles respectively. The volume V_{pump} of liquid discharged by the micropump can be estimated from the Hagen-Poiseuille formula [Equation (5.3)].

$$V_{pump} \approx \frac{\Delta P_m \pi R_o^4}{8\eta L_o} \tag{5.3}$$

Strictly speaking, (5.3) gives the rate of volume discharge. The η is the viscosity of liquid discharged and $L_o(R_o)$ is the length (inner radius) of the connecting tube. The micropump can be used for accurate delivery of small quantities of liquids and the design of microreactors. It can also be used in microfluidics (Bormashenko *et al.* 2010).

Example 5.1: Calculating the Rate of Volume Discharge into a Liquid Marble

A PVDF particle-stabilized water liquid marble (50 µL) having an effective surface tension of ~70 mN m^{-1} was connected to another water liquid marble of the same volume stabilized by lycopodium particles, having an effective surface tension of ~50 mN m^{-1}, by a tube of inner diameter 1 mm and length 0.5 cm. (a) Which liquid marble will lose liquid to the other? (b) Taking the viscosity of water as 10^{-3} Pa s and radius of the marbles as 2.3 mm, calculate the rate at which the discharging liquid marble will discharge liquid into the other.

Method

(a) Determine the Laplace pressures of the liquid marbles. The marble with the larger pressure will discharge liquid into the one with smaller pressure. (b) Combine Equations (5.1) and (5.3) and solve for the discharge volume rate V_{pump}.

Answer

1. From

$$\Delta P_m \sim \frac{2\gamma_{eff}}{R_m}$$

For PVDF-stabilized liquid marble (50 µL) with $\gamma_{eff} = 70$ mN m^{-1} = 70×10^{-3} N m^{-1}:

Volume of marble $V_m = \frac{4}{3}\pi R_m^2$

$$\Rightarrow R_m = \sqrt{\frac{3V_m}{4\pi}} \text{ where } V_m = 50 \times 10^{-9} \text{m}^3$$

$$= \sqrt{\frac{3 \times 50 \times 10^{-9} \text{ m}^3}{4\pi}} = 1.09 \times 10^{-4} \text{ m}$$

$$\Rightarrow \Delta P_m \sim \frac{2 \times 70 \times 10^{-3} \text{ N m}^{-1}}{1.09 \times 10^{-4} \text{ m}} = 1.28 \times 10^3 \text{ N m}^{-2}$$

For lycopodium-stabilized liquid marble (50 µL) with $\gamma_{eff} = 50$ mN m^{-1} = 50×10^{-3} N m^{-1}:

$$\Rightarrow \Delta P_m \sim \frac{2 \times 50 \times 10^{-3} \text{ N m}^{-1}}{1.09 \times 10^{-4} \text{ m}} = 9.17 \times 10^2 \text{ N m}^{-2}$$

Therefore because the pressure of the PVDF-stabilized marble is larger than that of the lycopodium-stabilized one, it will discharge liquid into it.

2. The combination of Equations (5.1) and (5.3) gives

$$V_{pump} \approx \frac{\pi R_o^4}{8\eta L_o} \times \frac{2|\Delta\gamma_{eff}|}{R_m}$$

$$\Rightarrow V_{pump} \approx \frac{\pi \times (5\times10^{-4})^4 \text{ m}^4}{8\times(10^{-3} \text{ N s m}^{-2})\times(5\times10^{-3} \text{ m})} \times \frac{2\times(70-50)\times10^{-3} \text{ N m}^{-1}}{2.3\times10^{-3} \text{ m}}$$

$$\approx 8.54\times10^{-8} \text{ m}^3 \text{ s}^{-1} \text{ or } 85.4 \text{ μL s}^{-1}$$

5.4 LIQUID MARBLES IN MINIATURIZED CHEMICAL PROCESSES

Miniaturized chemical processes are micrometre-scale processes. Unlike their macroscale counterparts, they use smaller volumes of reagents and as a result generate less waste. Because of the small volume of chemical reagents involved, miniaturized chemical processes are also cheaper, quicker and less harmful compared to their macroscale counterparts. In addition, conditions in a miniaturized chemical process can be easily controlled. Also, miniaturized chemical processes can be easily integrated in a digital device. Furthermore, they are very useful for high-throughput analyses and purifications in chemical and biological processes like drug discovery, DNA analysis, protein crystallization, and molecular synthesis. Due to their small size, non-wetting property, and the tendency to be easily manipulated by external forces like gravity, electric, and magnetic fields depending on the encapsulated liquid, liquid marbles provide optimal conditions for miniaturized chemical processes. For example, Dorvee *et al.* (2004) performed a precipitation reaction using liquid marble technology. A KI solution and $AgNO_3$ solution marbles, both stabilized by magnetic porous silicon particles, were forced to coalesce under the influence of a magnetic field, giving rise to a larger marble containing precipitates of AgI formed from the reaction of KI and $AgNO_3$. In a related study, Xue *et al.* (2010) showed that liquid marbles can be used as smart miniature reactors for different chemical reactions. Liquid marbles, stabilized by superparamagnetic fluorinated decyl polyhedral oligomeric silsesquioxane/Fe_2O_3 composite nanoparticles, containing hydrogen peroxide and bis(2,4,6-trichlorophenyl)oxalate and dye were used for the experiment. The particle coating rendered the liquid marbles magnetic and they were forced to coalesce using a magnetic field thereby initiating the chemiluminescence reaction. In a parallel experiment, a bicomponent liquid marble, *i.e.* containing more than one liquid type, *e.g.* oil-in-water or water-in-oil, was used to carry liquid drops containing the active ingredients in an immiscible fluid. Through gentle movement, actuated by a magnetic field, the liquid drops coalesced, leading to the chemiluminescence reaction. Photochemical reactions, nanoparticle synthesis, and acid-base reactions were also reported to be successful (Xue *et al.* 2010). In another experiment by Binks *et al.* (2017), mixed liquid marbles of water and cyclohexane, stabilized by fluorinated sericite clay particles, were used to miniaturize the so-called nylon-rope trick experiment. The cyclohexane marble containing sebacoyl chloride and water marble containing hexamethylene diamine were coalesced mechanically, initiating the reaction, and Nylon 6-10 was obtained.

Liquid marbles have also been used as microelectrochemical reactors where transportation of electric charges between liquid marbles of electrolyte solutions was demonstrated, leading to the construction of a Daniell cell whose potential $(1.085 \pm 0.004 \text{ V})$ is similar (1.1 V) to that of a Daniell cell [Figure 5.2 (upper image)] formed with solutions in beakers (Li *et al.* 2013). Because the potential generated by the cell is < 1.1 V, three cells were connected in series to generate a potential of ~3 V, sufficient to power an LED. The liquid marble Daniell cell was made up of 0.1 M $ZnSO_4$ marble (50 µL) and 0.1 M $CuSO_4$ marble (50 µL) with Zn (0.2 mm) and Cu (0.2 mm) electrodes immersed in the $ZnSO_4$ and $CuSO_4$ solutions, respectively, as two half-cells and an agar salt bridge. The electrolyte solutions were wrapped with PTFE powdered particles (size 100 µm) to obtain the marbles. The salt bridge was prepared by first heating a mixture of 3% agar in 1 M KNO_3 (w/v) followed by filling a polyethylene tube (i.d. 1 mm) with the solution and allowing to cool to room temperature. The ends of the salt bridge were then inserted into the $ZnSO_4$ and $CuSO_4$ liquid marbles, connecting them and forming a liquid marble Daniell cell [Figure 5.2 (lower image)].

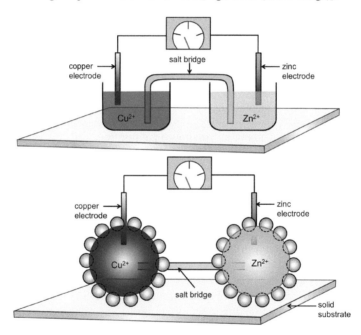

FIGURE 5.2   The upper part of the sketch describes a typical solutions-in-beaker Daniell cell and the lower part describes liquid marble Daniell cell. (Re-drawn from Li, M. *et al.*, *Chem. Eng. J.*, 97, 337–343, 2013.)

Particles adsorbed on the surfaces of liquid marbles are partially immersed in the liquid drop while protruding into the air phase. This provides a suitable platform for site-selective modification of the particles and can be used for the fabrication of Janus particles. Janus particles are particles with "solvent-loving" and "solvent-hating" portions and are important in emulsion formulation. The fabrication of Janus particles using miniature liquid marbles has been reported (Sheng *et al.* 2016). Dopamine

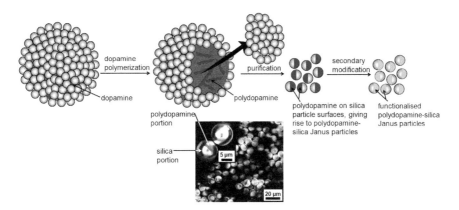

FIGURE 5.3 The upper image shows the schematic description of Janus particles synthesis using dopamine polymerization in liquid marbles. Stable liquid marbles are prepared and used as microreactors for dopamine polymerization. The polymerized dopamine selectively adheres to the silica surfaces in the aqueous phase leading to the formation of polydopamine-silica Janus particles which are purified and further modified *via* physical or chemical processes to obtain other Janus particle types. The lower part shows SEM images of polydopamine-silica Janus particles. (Re-drawn and taken from Sheng, Y. *et al.*, *Langmuir*, 32, 3122–3129, 2016.)

hydrochloride aqueous solution (4 g L^{-1}) was wrapped with superhydrophobic silica particles (size 10 μm) and injected with 20 mM isometric Tris solution (pH = 8.5), leading to the formation of dopamine monomers which polymerized rapidly into polydopamine in the liquid marbles. The polydopamine selectively adsorbed on the aqueous solution-particle surfaces of the superhydrophobic silica particles, giving rise to polydopamine-silica Janus particles as illustrated in [Figure 5.3 (upper image)]. After the Janus particle formation reaction was complete (12 h), the liquid marbles were immersed in ethanol where the particles desorbed from the drop surfaces. The particle suspension precipitated out rapidly while the excess reactants and polydopamine were removed as supernatant. Finally, the Janus particles were washed several times with ethanol and freeze-dried. The scanning electron microscope (SEM) image of the freeze-dried particles [Figure 5.3 (lower image)] clearly shows the Janus nature of the particles. The polydopamine portion is hydrophilic while the silica portion is superhydrophobic.

The Janus particles were further functionalized using physical or chemical processes as illustrated in Figure 5.4, leading to the formation of other Janus particle types. For example, the Janus particles were dispersed in ethanol solution followed by addition of CaCO$_3$ (size 20 nm) suspension and left for 12 h. The CaCO$_3$ particles adhered to the sticky polydopamine portion of the Janus particles, leading to the formation of CaCO$_3$-silica Janus particles which were filtered and freeze-dried. In a parallel modification experiment, the liquid marbles contents were withdrawn and injected with deionized water followed by addition of the CaCO$_3$ particle suspension and left in an air-sealed vial for 12 h. Thereafter, the marbles were dispersed in ethanol followed by particle filtration, washing with ethanol and freeze-drying.

Liquid marbles have also been used as microreactors for the miniaturized synthesis of nanocomposite. A schematic illustration for such a process is shown in Figure 5.5 for the synthesis of graphene/Ag nanocomposite. Thermally robust liquid

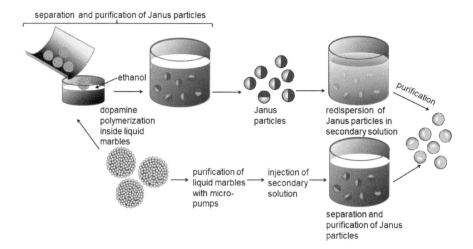

FIGURE 5.4 Schematic diagram for the modification of polydopamine-silica Janus parti-cles. Liquid marbles are dispersed in ethanol (upper part of the image), leading to the desorp-tion of the particles from the marbles surfaces. The particles are further modified to obtain other particles by re-dispersing them in a secondary liquid followed by purification. The lower part of the image shows an alternative modification route where the marble microreac-tor is purified by replacing the reacting solution with deionized water using a micropump. The marble is then injected with a secondary solution, leading to the functionalization of the polydopamine portion. Lastly, the functionalized Janus particles are separated and purified. (Re-drawn from Sheng, Y. *et al.*, *Langmuir*, 32, 3122–3129, 2016.)

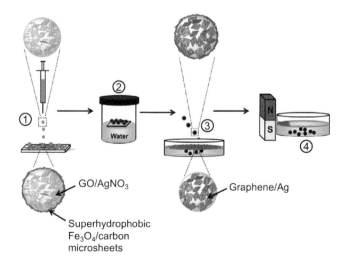

FIGURE 5.5 A schematic depiction of miniaturized synthesis of graphene/Ag nanopar-ticles using liquid marbles: (1) preparation of liquid marbles containing $GO/AgNO_3$ solution using superhydrophobic Fe_3O_4/carbon microsheets, (2) chemical reduction of $GO/AgNO_3$ to graphene/Ag nanocomposite, (3) separation, and (4) collection of the graphene/Ag nano-composite from the superhydrophobic microsheets. (From Chu, Y. *et al.*, *ACS Appl. Mater. Interfaces*, 6, 8378–8386, 2014.)

marbles, prepared by encapsulating ~5 µL of a mixture of 7–20 mg $AgNO_3$ and 14–21 mg ascorbic acid dissolved in 1 mL of graphite oxide (GO) with superhydrophobic Fe_3O_4/carbon microsheets, were used for the experiment (Chu *et al.* 2014). The liquid marbles were placed on a water bath (80°C) for 50 mins, where the mixture was reduced to graphene/Ag nanocomposite, followed by transferring to water at room temperature, where the superhydrophobic particles desorbed from the nanocomposite surfaces. The superhydrophobic Fe_3O_4/carbon microsheets were removed using a magnet, leaving the spherical graphene/Ag nanocomposite. Lastly, the nanocomposite was dialyzed in deionized water for 72 h to remove residual ascorbic acid. The graphene/Ag nanocomposite obtained exhibited one of the best catalytic characteristics for 4-nitroaniline reduction into *p*-phenylenediamine compared with other reported catalysts (Chu *et al.* 2014).

In a parallel study, a hydrophobic cyclomatrix polyphosphazene PZAF/Ag nanoparticle composite, prepared by one-step precipitation of polycondensation of hexachlorocyclotriphosphazene and 4,4′-(hexafluoroisopropylidene)diphenol, followed by in-situ reduction of $AgNO_3$, was used for the fabrication of catalytic liquid marbles (Wei *et al.* 2016). It was shown that the reduction of methylene blue in aqueous solution by $NaBH_4$ can be efficiently catalyzed in the catalytic liquid marbles irrespectively of the marble volume.

5.5 LIQUID MARBLES AS PRECURSORS OF PICKERING EMULSIONS

Emulsions are thermodynamically unstable mixtures of immiscible liquids in which one of the liquids is dispersed as microscopic drops in the bulk of the other. Emulsions are kinetically stabilized (*i.e.* prevented from separating out) using small amphiphilic organic molecules called surfactants, polymer molecules like proteins, and powdered particles alone or in combination with surfactants. Emulsions which are kinetically stabilized by powdered particles are known as Pickering emulsions, named after Professor Spencer Umfreville Pickering (1858–1920), who showed that it is possible to obtain stable emulsions using powdered particles based on the ideas of Professor Walter Ramsden (1869–1947). Conventional preparation of Pickering emulsions involves shearing powdered particles with the immiscible liquid mixture for a given period of time. Bormashenko *et al.* (2012) have demonstrated that Pickering emulsions can also be prepared from liquid marbles. Initially, water and glycerol liquid marbles were separately prepared from five different powdered particles (PTFE, PVDF, PE, carbon black, and lycopodium). Thereafter, the liquid marbles were immersed in various organic liquids to obtain the Pickering-like emulsions. It was reported (Bormashenko *et al.* 2012) that nonpolar oils like polydimethylsiloxane, toluene, xylene, and chlorinated solvents supported the formation of emulsions, whereas polar liquids like dimethylsufoxide, N,N-dimethylformamide, acetone, and ethanol did not. Using this method, one can overcome the difficulty associated with preparing millimetre-sized Pickering emulsions of narrow size distribution. However, this method may be time consuming considering the fact that the liquid marbles have to be prepared one-by-one before dispersing in the desired bulk continuous liquid phase.

5.6   LIQUID MARBLES AS MICROBIOREACTORS

Liquid marbles can also be used as microbioreactors for biological reactions and diagnostic assays. The advantages of a liquid marble microbioreactor are numerous. Liquid marble microbioreactors require relatively small amount of samples and reagents. Biohazard risks are limited because the powder coating encapsulates the sample and prevents contact with the supporting substrate. Marbles microbioreactor reactions can easily be controlled. Lastly, marbles are low in cost and can be easily disposed-off after the reaction. Human blood (grouping ABO and Rh) has been used to demonstrate the use of liquid marbles as microbioreactors (Arbatan *et al.* 2012b). Blood type assessment is indispensable in blood transfusion to prevent severe consequences of blood incompatibility. A schematic illustration of blood grouping test using liquid marble technology is shown in Figure 5.6 (Arbatan *et al.* 2012b).

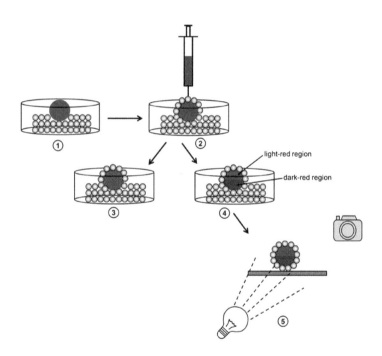

FIGURE 5.6   Schematic of blood marble preparation and its application in blood-type identification. (1) A drop (10 µL) of blood is deposited on hydrophobic powdered particles, leading to the formation of blood marble upon rolling. (2) Injection of antibody solution (10 µL) into the blood marble. (3) No phase separation occurs as the corresponding antigens are not present in the blood marble. (4) Phase separation occurs as the corresponding antigens are present in the blood marble due to agglutination reaction. (5) The blood marble in (4) is placed on a solid substrate, illuminated by white light to improve visualization of agglutination reaction and photographed for archiving. (Re-drawn from Arbatan, T. *et al.*, *Adv. Healthcare Mater.*, 1, 80–83, 2012b.)

In line with the illustration, a blood sample was used to prepare three blood marbles (each 30 µL) using precipitated $CaCO_3$ particles treated with stearic acid or PTFE particles. This was followed by the injection of 20 µL of antibody solutions (Anti-A, Anti-B, and Anti-D). Lastly, the ABO and Rh blood grouping was determined by checking whether or not haemagglutination reaction occurred in the blood marble microbioreactors. The occurrence of haemagglutination reaction was associated with a two-phase separation (Figure 5.7), light-red and dark-red colored regions, in the marble. The dark-red colored region resulted from the precipitation of agglutinated red blood cells to the bottom of the marble, indicating the presence of the corresponding antigen. By contrast, the absence of the corresponding antigen was

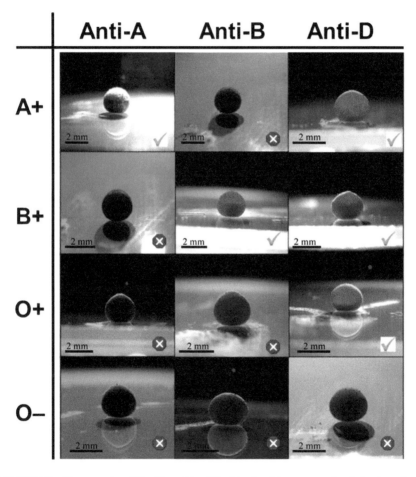

FIGURE 5.7 Photographs of liquid marbles (20 µL) microbioreactors after injection of antigens. Green ticks are placed on photos of marbles where phase separation due to agglutination reaction has occurred while red crosses are placed on those without phase separation. (From Arbatan, T. *et al.*, *Adv. Healthcare Mater.*, 1, 80–83, 2012b.)

indicated by the absence of phase separation in the marble. In addition to the low-cost, low-biohazard risks and disposability of the waste associated with the technique, no special medical facilities are needed for the test.

5.7 LIQUID MARBLES AS A MEANS OF CULTIVATING MICROORGANISMS AND CELLS

The particle coating on liquid marbles prevents the liquid content from direct contact with the external surfaces, but allows the free exchange of gases between the liquid and external environment. This property makes liquid marbles suitable microbiore-actors for cultivating microorganisms, bacteria, or cells. This has been demonstrated by Tian *et al.* (2013) using separate aqueous liquid marbles (20 µL), stabilized by Teflon powdered particles of diameter 100 µm, loaded with cultures of anaerobic *Lactococcus lactis cremoris* (aerotolerant anaerobe) and *Saccharomyces cerevisiae* (facultative aerobe). The proliferation of the microorganism cells in the liquid marbles upon incubation (at 30°C for 24 h) was different depending on their oxygen demand as expected. The liquid marbles were seen to provide a suitable environment for the growth of the aerobic microorganism compared to the anaerobic one. The different proliferation rate of the microorganism cells in the marbles was linked to the presence and free exchange of oxygen and carbon dioxide across the marble porous coating. Also, the small volume of the marbles gives them the inherent merit of having a large air-to-liquid ratio necessary for a faster proliferation of the aerobic microorganism as seen compared to bulk liquid cultures, *e.g.* using the conventional McCartney bottles.

In other experiments, liquid marbles have been used as three-dimensional micro-bioreactors for the production of three-dimensional cell structures. For example, cancer cell spheroids (CCSs) cultured in vitro are known to closely resemble the in vivo physiology of tumours compared with the two-dimensional cell cultures. As a result, the in vitro cultured CCSs are generally used as models in studying the physiology of tumours and several methods of producing them have been developed.

One commonly used method is the hanging drop method. Unfortunately, the method is onerous, requiring a careful drop manipulation and small drop size. Liquid marbles, stabilized by inert PTFE particles, have been used as a novel alternative to the hanging drop for the formation of CCSs as illustrated in Figure 5.8a, b (Arbatan *et al.* 2012a). Unlike the hanging drop method, with a drop size limit, the relatively large size of liquid marbles leads to the formation of a relatively large number of cell aggregates and three-dimensional cell structures. In addition, the marble is viable for this experiment for three reasons. The confined marble liquid core ensures close cell contacts and hence greater cell aggregation. The porous particle shell allows the free exchange of O_2 and CO_2 between the cultured medium in the marble and the environment. Lastly, the poor adhesive tendency of the cells to the hydrophobic particle shell promotes their suspension in the cultured medium. A typical CCSs formation experiment involves placing a drop (100–400 µL) of Hep G2 cells suspension encapsulated with PTFE powder (size = 100 µm) on the surface of bulk water in a closed petri dish, followed by incubation. The water ensures saturation of the atmosphere with water vapor to minimize evaporation of the marble

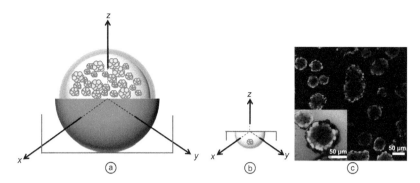

FIGURE 5.8 Illustration of cancer cell spheroid formation in (a) a liquid marble and (b) a hanging drop. Notice that the particle coating on the liquid marble is not shown. Compared with the hanging drop method which is limited by drop size, marbles can contain relatively large amount of liquid making it possible for large number of cell aggregates to form. (c) Confocal scanning microscope image of Hep G2 cells aggregates formed in liquid marbles ten days after incubation. The cells were stained with Calcein AM (green) prior to viewing. The formation of dark lining around the cell aggregates is due to the transition of cell aggregates into tumour spheroid. (From Arbatan, T. *et al.*, *Adv. Healthcare Mater.*, 1, 467–469, 2012a.)

aqueous content. Finally, cell aggregation is monitored by examining extracted marble content *via* confocal microscopy. Cell aggregation and formation of three-dimensional cell aggregates are observed after one day and ten days, respectively. Three-dimensional aggregates (tumour spheroids) as large as 100 μm have been obtained as shown in the image in Figure 5.8c.

Using liquid marble microbioreactor for CCSs formation has several advantages over the hanging drop method. First, the CCSs form within a relatively short period of time. Second, numerous CCSs are produced from a single marble microbioreactor. Third, the method is simple and does not require sophisticated instrumentation. Fourth, human interference is low once the marble is prepared and incubated. Fifth, the flexible nature of the marble cell allows the extraction and replenishment of its content. Last, the cost of producing a liquid marble bioreactor is less compared with the hanging drop microbioreactors (Arbatan *et al.* 2012a).

In another study by Vadivelu *et al.* (2015), the three-dimensional culturing of olfactory ensheathing cells (OECs), using floating aqueous liquid marbles (10 μL), stabilized by PTFE particles of average size 1 μm, was reported. The OECs are glial cells of the olfactory nervous system having unique growth-promoting properties that can dramatically increase survival and axonal regeneration of the central nervous neurons. Transplantation of OECs is being trialled for repair of paralyzed spinal cord. Unfortunately, the investigation of OEC behavior in a multicellular environment has been hindered by lack of suitable three-dimensional cell culture models. The study (Vadivelu *et al.* 2015) showed that the floating marbles allowed the OECs to freely associate and interact to form OEC spheroids of uniform shapes and sizes. The floating marbles were also used to co-culture the OECs with Schwann cells and astroyctes so as to investigate how the OECs and other cell types associate and interact while forming complex cell structures.

Aqueous liquid marbles have also been used for cryopreservation of mammalian cells as an alternative to conventional methods, which often require the use of cryopreservative agents with associative degree of cell toxicity (Serrano *et al.* 2015). The marbles were stabilized with hydrophobic PTFE particles. The marbles were loaded with the mammalian cells and frozen. The cells were recovered upon thawing the marbles without a significant change to their cellular parameters like adhesion, morphology, viability, proliferation, and cell cycle.

In another demonstration experiment, a drop of murine embryonic stem cell (ESC) (Oct4B2-ESC) suspension was deposited onto a powdered bed of PTFE particles leading to the formation of its liquid marble (Sarvi *et al.* 2014). The Oct4B2-ESC formed suspended embryoid bodies (EBs) *via* aggregation with relatively uniform size and shape with some elements of cardiogenesis, in the marble within three days, indicative that the marble is a favorable, microenvironment to induce EBs formation as well as their cardiogenesis.

5.8   DRUG SCREENING BASED ON LIQUID MARBLE TECHNOLOGY

The process of identifying and optimizing potential drugs that leads to the selection of a candidate drug for clinical trials is known as drug screening. The process involves screening a large number of drug-like chemical compounds for a specific biological activity in high-throughput screening assays. In addition to the culturing of microorganisms and the preservation of cells in liquid marbles, the use of liquid marbles for high-throughput drug screening of anchorage-dependent cells that require a physical support to adhere and proliferate has been demonstrated (Figure 5.9a, Oliveira *et al.* 2014). Drops (20 µL) of aqueous suspension containing anchorage-dependent L929 cells and poly(L-lactic acid) microparticles (size 20–100 µm), whose surfaces have been modified by plasma treatment with collagen, were encapsulated with hydrophobic diatomaceous earth (hydrophobized by fluorosilanization), leading to the formation of liquid marbles. Prior to the experiment, it was verified that the surface modified poly(L-lactic acid) microparticles and the hydrophobic diatomaceous earth are nontoxic to the cells. The microparticles provided anchorage sites necessary for cell adhesion that leads to proliferation. Diatomaceous earth, shown in Figure 5.9b, is the exoskeletons (unique silica-based microskeletons with nanotextures on their surfaces) of unicellular microalgae that remain and sediment on the bottom of lakes or sea upon the death of the microalgae. The nanotextures impart special features to the particles including water-repellency especially when further chemically modified.

The liquid marbles were incubated (37°C) for 24 h where cell aggregation and proliferation were observed. Because the liquid content of liquid marbles can be removed or increased without collapsing them, Fe^{3+} from a solution of $FeCl_3.6H_2O$ (6–18 mM) was injected into the marbles and incubated for another 24 h. The Fe^{3+} served as a cytotoxic agent or drug to be screened. Iron is very important in many biological processes like DNA synthesis, erythropoiesis, and electron and oxygen transportation, but it is also toxic because in the presence of air it catalyzes reactions

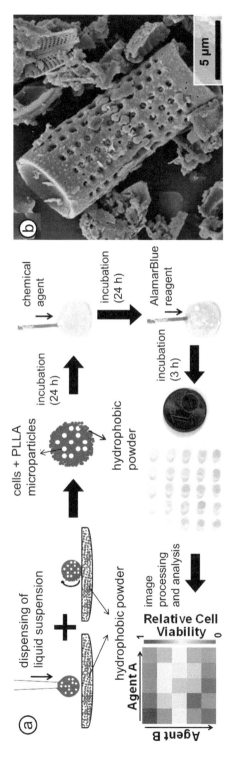

FIGURE 5.9 (a) Schematic description of high-throughput drug screening using liquid marbles containing anchorage-dependent cells and PLLA microparticles, showing stages of marble formation, marble incubation, chemical agent (drug) injection, and cell viability determination. (b) SEM image of diatomaceous earth prior to hydrophobization. (From Oliveira, N.M. *et al.*, *Adv. Healthcare Mater.*, 4, 264–270, 2014.)

that lead to the formation of reactive oxygen species and other reactive radicals like the hydroxyl radicals.

Following incubation, the viability of cells in the marbles containing Fe^{3+} was measured by color change using AlmarBlue reagent or MTS (3-(4,5-dimethylthiazol-2-yl)-5-(3-carboxymethoxyphenyl)-2-(4-sulfophenyl)-2H-tetrazolium) colorimetric assay or DNA quantification and compared with those without the metal ion. In AlmarBlue colorimetric assay, a resazurin-based product (blue) is reduced to resofurin (red) in the presence of viable cells. Similarly, in the MTS assay, a tetrazolium salt is reduced to a formazan product with an associated color change of yellow to brown in the presence of viable cells. The intensity of the resultant color and DNA amount ratio (as compared with a marble containing phosphate buffer saline) was seen to decrease with increasing Fe^{3+} concentration as expected, indicative that cell viability decreases with Fe^{3+} concentration and that Fe^{3+} toxicity is concentration dependent. This shows that liquid marbles can be used for high-throughput drug screening of anchorage-dependent cells as well as other cell types.

5.9 LIQUID MARBLES AS PRECURSORS OF NOVEL MATERIALS

Liquid marbles are precursors of the so-called powdered liquids which are also known as dry liquid powders. Powdered liquids are liquid-in-air materials (*cf.* liquid foams which are air-in-liquid materials). A powdered liquid material is a novel free-flowing powder composed of small (diameter 50–400 μm) liquid drops coated with powdered particles of low surface energy which poorly wet them. The liquid drops may be water (Binks and Murakami 2006, Binks *et al.* 2010), oils (Binks *et al.* 2014, 2015), or emulsions (Murakami *et al.* 2012, Binks and Tyowua 2016b). Dry water, oil, and emulsions are the terms used to describe powdered liquids if the liquid drops coated with particles are water, oil, and emulsions, respectively. The liquid drops in powdered liquids are usually non-spherical. This is due to the jamming of particles on their surfaces. This gives the drop surfaces solid-like properties which make relaxation to a spherical geometry impossible. It is important to note that the properties (*e.g.* contact angle) of the powdered particles used in the stabilization of powdered liquids are the same as those required for the precursor liquid marble. For example, hydrophobic particles are required for the stabilization of water liquid marbles and powdered water (Murakami and Bismarck 2010). Similarly, oleophobic particles are required for the stabilization of oil liquid marbles and powdered oils (Binks *et al.* 2014, 2015). In the case of powdered emulsions, the continuous phase of the emulsions needs to have low affinity for the particles encapsulating it (Carter *et al.* 2011, Murakami *et al.* 2012). Because the particles protect the liquid drops against coalescence, powdered liquids are stable against phase separation for several years. The encapsulated liquid may be released on application of mechanical stress or *via* evaporation if its vapor pressure is relatively high. This property makes them suitable candidates in drug delivery and cosmetics.

The phase inversion of powdered liquid materials yields foams. This is achieved, mainly, by increasing the volume fraction of the liquid phase at fixed air to particle ratio (Binks *et al.* 2014, 2015). In the case of powdered emulsions, phase inversion yields a paste (Carter *et al.* 2011). In some cases, it is possible to phase invert a foam to a powdered liquid. By either changing the particle hydrophobicity progressively (fixed air to water ratio) or by changing the air to water ratio (fixed particle wettability), a particle-stabilized aqueous foam was inverted to powdered water (Binks and Murakami 2006).

5.9.1 Dry Water and Applications

Dry water is a free-flowing powder containing tiny water drops (of the order of 50 μm) coated with a network of hydrophobic powdered particles which armour them against coalescence.

Dry water is prepared by shearing hydrophobic powdered particles and water at high speed, *e.g.* 19 000 rpm for a period of 1 to 2 min, in the presence of air as described schematically in Figure 5.10a. A photograph of a typical dry water sample is shown in Figure 5.10b passing freely through a funnel. The Cryogenic scanning electron microscope (Cryo-SEM) image of one of the powder grain in Figure 5.10b is shown in Figure 5.10c with the particles on its surfaces.

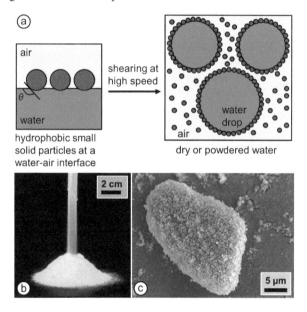

FIGURE 5.10 (a) Schematic of hydrophobic powdered particles at a water-air interface, making a contact angle θ of more than 90° with it and the corresponding dry or powdered water formed upon shearing at high speed. (b) Photograph of dry water containing water drops coated with hydrophobic silica particles passing freely through a glass funnel. (c) Cryo-SEM image of one of the powder grain in (b) showing silica particles on its surfaces. (From Binks, B.P. *et al.*, *Soft Matter*, 6, 126–135, 2010.)

Dry water is also considered as a dispersion of particle coated-water drops in air. The particle coated-drops are stable against coalescence and phase separation for several years, but their diffusion-controlled evaporation depends on the ambient conditions with a higher temperature enhancing it. However, in sealed vial, dry water is stable to evaporation for several years. Evaporation of the water drops is also lowered by adding thickening agents to the water phase prior to dry water preparation. Dry water phase inverts to aqueous liquid foam.

Dry water has been used for several applications. For example, its highly distributed gas-water interface has been used to greatly enhance the kinetics of gas-water heterogeneous catalytic hydrogenation (Carter *et al.* 2010). Dry water has been used as gas hydrate to store gases like CH_4, CO_2, and Kr (Wang *et al.* 2008, Carter *et al.* 2009). Gas hydrates or gas clathrates are nonstoichiometric crystalline inclusion solid compounds made up of a hydrogen-bonded water lattice which trap small molecules, like CH_4, O_2, CO_2, H_2S, H_2, and N_2, within polyhedral cavities. They occur naturally in large quantities. These water-based crystalline solids resemble ice physically but are chemically distinct from ice given the presence of trapped gaseous molecules in the cavities of the frozen water molecules. Gas hydrates have been used for the capturing, separation, storage, and transportation of gases. One limitation of gas hydrates is their slow formation owing to the small gas-solid or gas-liquid interface involved. To surmount this problem, gas hydrates have been prepared using dry water by taking advantage of the high interfacial contact between the water drops and air. Because of the high interfacial contact between the water drops and air, gas diffusion into the hydrate structure is greatly enhanced compared to the bulk water or ice structure leading to an increase in the kinetics of hydrates formation. The stability of the dry water powder to evaporation and its recyclability in terms of gas hydrate formation is greatly enhanced when a gelling agent like gellan gum is added to it (Carter *et al.* 2009).

Dry water has also been used in the cosmetic and pharmaceutical industries for microencapsulation of active ingredients (Zamyatin *et al.* 2007). A dry water cosmetic composition that changes color upon shearing on a surface has been reported. Suitable organic or inorganic pigments were added to the water phase before shearing with hydrophobic fumed silica particles at high speed to obtain the dry water. The dry water powder was meant to be used as lipsticks, eye shadows, foundations, mascaras, eye liners, and other related applications (Zamyatin *et al.* 2007).

5.9.2 POWDERED OILS AND APPLICATIONS

Powdered oils (Figure 5.11) are free-flowing powders containing tiny oil drops (≤ 400 μm) coated with a network of oleophobic small solid particles which armour them against coalescence. They are prepared by shearing drops of the necessary oil with oleophobic powdered particles at low speed, for a given period, up to a critical oil to particle ratio. Above the critical oil to particle ratio, powdered oils phase invert to oil foams. Powdered oils are considered as a dispersion of particle coated-oil drops in air in which the oil drops are stable against coalescence, phase separation, and diffusion-controlled evaporation (unless very volatile) for several years. Different oils have been used in the preparation of powdered oils (Binks *et al.* 2014, 2015). Using fluorinated clay sericite and fluorinated ZnO particles, powdered oils were prepared from different oils ranging from non-polar linear and branched mineral oils (alkanes), polar (vegetable) oils like jojoba oil, and liquids like glycerol and other perfumery liquids like limonene for different cosmetic and pharmaceutical applications.

FIGURE 5.11 (inset) Photograph of powdered squalane, obtained by coating squalane drops with oleophobic fluorinated ZnO particles, passing freely onto a Pyrex glass, and a corresponding Cryo-SEM image of the powder, showing the particles on their surfaces. (From Binks, B.P. *et al.*, *ACS Appl. Mater. Interfaces*, 7, 14328–14337, 2015.)

5.9.3 POWDERED EMULSIONS AND APPLICATIONS

As mentioned previously, emulsions are a dispersion of microscopic drops of one liquid in the bulk of another liquid in the presence of a suitable emulsifier, in which the two liquids are immiscible at ordinary conditions (Schramm 2005). When oil drops are dispersed in the bulk of a water continuous phase the emulsion is said to be of oil-in-water (o/w) type and water-in-oil (w/o) emulsion when water drops are dispersed in the bulk of an oil continuous phase (Schramm 2005). Certain oil pairs are also immiscible and can be sheared in the presence of a suitable emulsifier to obtain oil-in-oil (o/o) emulsions (Binks and Tyowua 2016a). The o/w, w/o, and o/o emulsions are known as simple emulsions. Sometimes, a simple emulsion is in turn dispersed in a bulk liquid continuous phase, in the presence of a suitable emulsifier, giving rise to the so-called complex or multiple emulsions (Aserin 2008). For example, multiple oil-in-water-in-oil (o/w/o) and water-in-oil-in-water (w/o/w) emulsions are formed when simple o/w and w/o emulsions are dispersed in the bulk of an oil and water continuous phases, respectively. Similarly, multiple oil-in-oil-in-oil (o/o/o) emulsions are obtained when simple o/o emulsions are dispersed in a bulk oil continuous phase, in the presence of a suitable emulsifier (Tyowua *et al.* 2017).

Powdered emulsions are obtained by wrapping drops of kinetically stable emulsion systems (that is, emulsions stable to creaming, sedimentation, coalescence and phase separation), with powdered particles that poorly wet the bulk continuous phase. This has been illustrated by Murakami *et al.* (2012) where *n*-dodecane-in-water emulsion, diluted with water, stabilized by partially hydrophobic fumed silica particles was aerated rapidly in the presence of very hydrophobic fumed silica particles to obtain powdered *n*-dodecane-in-water emulsion as depicted in the sketch of Figure 5.12a. The resultant powdered emulsion is tri-phasic, consisting of *n*-dodecane drops dispersed in water globules themselves dispersed in air, with very hydrophobic fumed silica particles wrapping the air-water surfaces as shown

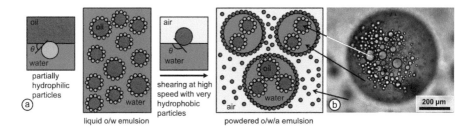

FIGURE 5.12 (a) Schematic description for the preparation of o/w/a materials. First, an o/w emulsion is stabilized by partially hydrophilic particles (green) followed by shearing at high speed in the presence of very hydrophobic particles (purple). The o/w/a material is composed of oil drops dispersed in water globules themselves dispersed in air, with the very hydrophobic particles coating the air-water surfaces. The excess unadsorbed particles retain their powdery nature. (b) Optical microscope image of *n*-dodecane-in-water-in-air material prepared by shearing (6 000 rpm) diluted *n*-dodecane-in-water emulsion in the presence of very hydrophobic particles, showing the water globule with the oil drops in them. (From Murakami, R. *et al.*, *Adv. Mater.*, 24, 767–771, 2012.)

in the optical microscope image in Figure 5.12b. As a result, the powdered material is described as oil-in-water-in-air (o/w/a) emulsion. The powdered material contains about 70 wt % of the precursor emulsion. In sealed vessel, the powdered material remained stable to coalescence and phase separation for over six months.

In another study, o/w/a materials were prepared by encapsulating a kinetically stable o/w emulsion stabilized by branched copolymer surfactants with hydrophobic fumed silica particles (Carter *et al.* 2011). The preparation of powdered w/o emulsions has also been reported (Binks and Tyowua 2016b). Kinetically stable water-in-vegetable oil and water-in-squalane emulsions stabilized by partially hydrophobic fumed silica particles were gently sheared in the presence of oleophobic fluorinated sericite particles. The resultant powders were triphasic, containing water drops dispersed in oil globules themselves dispersed in air and were also called water-in-oil-in-air (w/o/a) emulsions, with fluorinated sericite particles coating the air-oil surfaces. The powdered material contained about 80 wt% of the precursor emulsions and was stable to phase separation for over a year but exude oil and water upon shearing on a solid substrate. Above a critical w/o emulsion to fluorinated sericite particle ratio, the powdered materials invert to an emulsion paste, made up of air bubbles and water drops themselves dispersed in oil.

In the cosmetic and pharmaceutical industries, powdered emulsions can be used as carriers of active ingredients to the skin surface. They can also be used for the storage and transportation of harmful substances.

5.10 CONCLUSION

This chapter is the pinnacle of this book. It describes the various applications of liquid marbles with related limitations, where applications in microfluidics, environmental science, miniature chemical processes, medicine, and cosmetic and pharmaceutical industries are thoroughly presented. These applications are based on the non-wetting property and other behavior of liquid marbles.

EXERCISES

DISCUSSION QUESTIONS

Question 1

Write a short essay on the applications of liquid marbles in the following areas:

a. microfluidics,
b. environmental science,
c. miniature chemical processes, and
d. precursors of novel materials.

Question 2

Highlight the various applications of powdered liquid materials.

NUMERICAL QUESTION

Question 1

An aqueous liquid marble (40 µL) of effective surface tension 68 mN m^{-1} was connected to an oil liquid marble (25 µL) of effective surface tension 28 mN m^{-1} by a tube of inner diameter 0.8 mm and length 1 cm. Which marble will lose liquid to the other? At what rate will the discharging marble discharge liquid into the other? (Take the viscosity of water and that of the oil as 10^{-3} Pa s and 46.3 mPa s, respectively.)

FURTHER READING

McHale, G. and M. I. Newton. "Liquid Marbles: Topical Context within Soft Matter and Recent Progress." *Soft Matter* 11 (13) (2015): 2530–46.

Oliveira, N.M., R.L. Reis and J.F. Mano. "The Potential of Liquid Marbles for Biomedical Applications: A Critical Review." *Adv. Healthcare Mater.* 19 (6) (2017): 1700192.

REFERENCES

Arbatan, T., A. Al-Abboodi, F. Sarvi and P.P.Y. Chan. "Tumor inside a Peal Drop." *Adv. Healthcare Mater.* 1 (4) (2012a): 467–69.

Arbatan, T., L. Li, J. Tian and W. Shen. "Microreactors: Liquid Marbles as Micro-Bioreactors for Rapid Blood Typing." *Adv. Healthcare Mater.* 1 (1) (2012b): 80–83.

Aserin, A. (Ed). *Multiple Emulsions Technology and Applications.* New York: John Wiley & Sons, 2008.

Aussillous, P. and D. Quéré. "Liquid Marbles." *Nature* 411 (2001): 924–27.

Binks, B.P. and A.T. Tyowua. "Oil-in-Oil Emulsions Stabilized Solely by Solid Particles." *Soft Matter* 12 (3) (2016a): 876–87.

Binks, B.P. and A.T. Tyowua. "Particle-Stabilized Powdered Water-in-Oil Emulsions." *Langmuir* 32 (13) (2016b): 3110–15.

Binks, B.P. and R. Murakami. "Phase Inversion of Particle-Stabilized Materials from Foams to Dry Water." *Nat. Mater.* 5 (2006): 865–69.

Binks, B.P., A.J. Johnson and J.A. Rodrigues. "Inversion of 'Dry Water' to Aqueous Foam on Addition of Surfactant." *Soft Matter* 6 (1) (2010): 126–35.

Binks, B.P., J.M. Mooney and A.T. Tyowua. "Mixed Liquid Marbles: Preparation, Properties and Applications." *Colloids Surf. A* (Submitted) (2017).

Binks, B.P., S.K. Johnston, T. Sekine and A.T. Tyowua. "Particles at Oil-Air Surfaces: Powdered Oil, Liquid Oil Marbles, and Oil Foam." *ACS Appl. Mater. Interfaces* 7 (26) (2015): 14328–37.

Binks, B.P., T. Sekine and A.T. Tyowua. "Dry Oil Powders and Oil Foams Stabilized by Fluorinated Clay Platelet Particles." *Soft Matter* 10 (4) (2014): 578–89.

Bormashenko, E. and A. Musin. "Revealing of Water Surface Pollution by Liquid Marbles." *Appl. Surf. Sci.* 255 (2009): 6429–31.

Bormashenko, E., R. Balter and D. Aurbach. "Micropump Based on Liquid Marbles." *Appl. Phys. Lett.* 97 (9) (2010): 091908-08-2.

Bormashenko, E., R. Pogreb and A. Musin. "Stable Water and Glycerol Marbles Immersed in Organic Liquids: From Liquid Marbles to Pickering-Like Emulsions." *J. Colloid Interface Sci.* 366 (1) (2012): 196–99.

Bormashenko, E., R. Pogreb, Y. Bormashenko, A. Musin and T. Stein. "New Investigations on Ferrofluidics: Ferrofluidic Marbles and Magnetic-Field-Driven Drops on Superhydrophobic Surfaces." *Langmuir* 24 (21) (2008): 12119–22.

Carter, B.O., D.J. Adams and A.I. Cooper. "Pausing a Stir: Heterogeneous Catalysis in 'Dry Water'." *Green Chem.* 12 (5) (2010): 783–85.

Carter, B.O., J.V.M. Weaver, W. Wang, D.G. Spiller, D.J. Adams and A.I. Cooper. "Microencapsulation Using an Oil-in-Water-in-Air 'Dry Water Emulsion'." *Chem. Commun.* 47 (29) (2011): 8253–55.

Carter, B.O., W. Wang, D.J. Adams and A.I. Cooper. "Gas Storage in "Dry Water" and "Dry Gel" Clathrates." *Langmuir* 26 (5) (2009): 3186–93.

Chu, Y., Z. Wang and Q. Pan. "Constructing Robust Liquid Marbles for Miniaturized Synthesis of Graphene/Ag Nanocomposite." *ACS Appl. Mater. Interfaces* 6 (2014): 8378–86.

Dorvee, J.R., A.M. Derfus, S.N. Bhatia and M.J. Sailor. "Manipulation of Liquid Droplets Using Amphiphilic, Magnetic One-Dimensional Photonic Crystal Chaperones." *Nat. Mater.* 3 (12) (2004): 896–99.

Fujii, S., S. Kameyama, S.P. Armes, D. Dupin, M. Suzaki and Y. Nakamura. "pH-Responsive Liquid Marbles Stabilized with Poly(2-vinylpyridine) Particles." *Soft Matter* 6 (3) (2010): 635–40.

Li, M., J. Tian, L. Li, A. Liu and W. Shen. "Charge Transport between Liquid Marbles." *Chem. Eng. J.* 97 (2013): 337–43.

Murakami, R. and A. Bismarck. "Particle-Stabilized Materials: Dry Oils and (Polymerized) Non-Aqueous Foams." *Adv. Funct. Mater.* 20 (5) (2010): 732–37.

Murakami, R., H. Moriyama, M. Yamamoto, B.P. Binks and A. Rocher. "Particle Stabilization of Oil-in-Water-in-Air Materials: Powdered Emulsions." *Adv. Mater.* 24 (6) (2012): 767–71.

Oliveira, N.M., C.R. Correia, R.L. Reis and J.F. Mano. "Liquid Marbles for High-Throughput Biological Screening of Anchorage-Dependent Cells." *Adv. Healthcare Mater.* 4 (2) (2014): 264–70.

Sarvi, F., S.K. Jain, T. Arbatan *et al.* "Cardiogenesis of Embryonic Stem Cells with Liquid Marble Micro-Bioreactor." *Adv. Healthcare Mater.* 4 (1) (2014): 77–86.

Schramm, L.L. *Emulsions, Foams, and Suspensions: Fundamentals and Applications.* Weinheim, Germany: Wiley, 2005.

Serrano, M.C., S. Nardecchia, M.C. Gutierrez, M.L. Ferrer and F. del Monte. "Mammalian Cell Cryopreservation by Using Liquid Marbles." *ACS Appl. Mater. Interfaces* 7 (2015): 3854–60.

Sheng, Y., G. Sun and T. Ngai. "Dopamine Polymerization in Liquid Marbles: A General Route to Janus Particles Synthesis." *Langmuir* 32 (13) (2016): 3122–29.

Tian, J., N. Fu, X.D. Chen and W. Shen. "Respirable Liquid Marble for the Cultivation of Microorganisms." *J. Colloid Surf. B* 106 (2013): 187–90.

Tian, J., T. Arbatan, X. Li and W. Shen. "Liquid Marble for Gas Sensing." *Chem. Commun.* 46 (26) (2010a): 4734–36.

Tian, J., T. Arbatan, X. Li and W. Shen. "Porous Liquid Marble Shell Offers Possibilities for Gas Detection and Gas Reactions." *Chem. Eng. J.* 165 (1) (2010b): 347–53.

Tyowua, A.T., S.G. Yiase and B.P. Binks. "Double Oil-in-Oil-in-Oil Emulsions Stabilized Solely by Particles." *J. Colloid Interface Sci.* 488 (2017): 127–34.

Vadivelu, R.K., C.H. Ooi, R.-Q. Yao *et al.* "Generation of Three-Dimensional Multiple Spheroid Model of Olfactory Ensheating Cells Using Floating Liquid Marbles." *Scientific Reports* 5 (2015): 15083.

Wang, W., C.L. Bray, D.J. Adams and A.I. Cooper. "Methane Storage in Dry Water Gas Hydrates." *J. Am. Chem. Soc.* 130 (35) (2008): 11608–609.

Wei, W., R. Lu, W. Ye *et al.* "Liquid Marbles Stabilized by Fluorine-Bearing Cyclomatrix Polyphosphazene Particles and Their Application as High-Efficiency Miniature Reactors." *Langmuir* 32 (7) (2016): 1707–15.

Xue, Y., H. Wang, Y. Zhao *et al.* "Magnetic Liquid Marbles: A 'Precise' Miniature Reactor." *Adv. Mater.* 22 (43) (2010): 4814–18.

Zamyatin, T., I.D. Sandewick, J.G. Russ and S.K. Jabush. "Dry Water Cosmetic Compositions That Change Color Upon Application." U.S. Patent 0218024 A1, 2007.

Answers to Numerical Questions

CHAPTER 1

Q1b) 0.2 MPa
Q2b) 7.2 nm^3
Q3)   10.6 mN m^{-1}
Q4)   71.8 mN m^{-1}

CHAPTER 2

Q1b) 0°
Q1c) 42.2°
Q2a) (i) 55°, (ii) 80°, (iii) 62° and (iv) 66°

When θ (experimental) < 90°, the calculated angles are in good agreement to it. By contrast, when θ (experimental) > 90°, the calculated angles do not agree with it. This means that the formula for calculating contact angles work well only when the angles are less than 90°.

Q2b) Because F_g(0.98 μN) > F_c(0.69 μN) the raindrop will not remain stuck to the vertical window pane.

Q3)

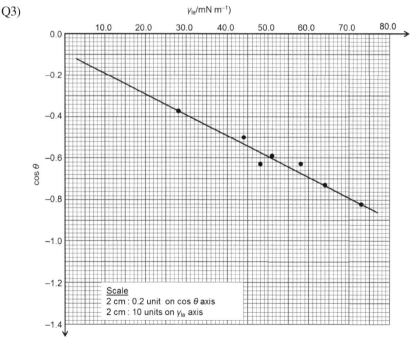

CHAPTER 3

Q1a) 2.27 mm

Q1c) 0.80 N

Q3c) The film thickness is 78.4 μm while the self-propelling force is 213.9 μN

CHAPTER 4

Q1a) The length of particle protrusion into the air phase is 3.5 μm while that into the drop phase is 0.5 μm. The drop diameter is 4.57 mm.

Q2a) 63.5 mN m^{-1} or 71.9 mN m^{-1}

Q2b) 72.7 mN m^{-1}

CHAPTER 5

Q1) Because the pressure (1390 N m^{-2}) in the 40 μL aqueous marble is larger than that (727 N m^{-2}) in the 20 μL oil one, the aqueous marble will discharge liquid into the oil one. The rate of volume discharge will be 10.7 mL s^{-1}.

Index

Note: Page numbers in italic and bold refer to figures and tables, respectively.